Edwin Ruston

Floral talks

Edwin Ruston

Floral talks

ISBN/EAN: 9783337271787

Printed in Europe, USA, Canada, Australia, Japan

Cover: Foto ©berggeist007 / pixelio.de

More available books at **www.hansebooks.com**

A MANUAL OF FLORICULTURE

BY

EDWIN RUSTON.

NEW YORK:
W. N. SWETT & CO., Publishers
28 Reade Street.
1892.

INTRODUCTION.

What recreation will afford more pleasure, involving a healthful exercise, than the cultivation of flowers? It is an exhilarating, interesting and ennobling work—so full of fascination—and when once begun it is likely to be continued. The love of flowers increases with all admirers of the beautiful as they are initiated into the mysteries of their culture and habits. Beauty is persistent and progressive, and when it is once imbued into the mind it rarely loses its hold. It is possible to make the poorest habitation charming by the judicious use of flowers, while a few radiant buds and blossoms may cheer a sad and weary soul like a sunbeam in an unlighted room, like the smiles on faces we love, like the soft speech of hearts uttered by faithful friends.

Every year shows a marked advance in the floral world, but to enumerate and describe all the various causes from which plants fail, would require a large and costly work, and this would require an annual revision to keep apace with the times, for new plant enemies, as well as plants are being discovered every year.

In this treatise it has been the writer's aim to give in a condensed practical shape, sufficient knowledge regarding the habits and requirements of such plants as may be cared for by anyone taking an interest in them with a small outlay of time and money, and with pleasure and satisfaction to the grower. It is hoped that the hints and suggestions offered in the following pages will be of much benefit to all lovers of flowers, and that the publication of this work may assist in the promotion of floriculture to the satisfaction and pleasure of many households.

E. RUSTON.

SYRACUSE, N. Y., April 1, 1892.

FLORAL TALKS.

There is nothing that will afford so much beauty and pleasure, at so little expense, as a window full of flowers. They are emblems of refinement, purity and happiness, and their good in the community is by no means small. The exercise involved in the care of plants is restful, because it is a change from the ordinary work. It is one of the greatest moral and physical health-giving recreations anyone can follow. It is educational, because whoever grows plants for the love of them is sure to become interested in their growth, and when once interested it is likely to be continued, for it is so full of fascination, and they are such wonderful teachers of nature. Dear reader, remember this, wherever you go: if you find flowers growing you will most generally find kind hearts and hospitality.

It is my desire to tell how I have grown such plants as I write about, with satisfaction to myself, and to help those who desire to cultivate them; and especially those who would grow flowers, but are afraid to for fear of failure, because of a lack of knowledge as to their cultivation.

Let me first say, however, that there are no set of rules which, if closely followed, will always lead to success in the culture of flowers. The treatment given one plant may not at all answer for another, therefore we must carefully study our plants, and in order to study them properly, one should love them, if we would know how to care for them successfully.

What Flowers to Grow in the House.

Many have been known to fail in floral culture simply because they commence at the top of the ladder, so to speak, instead of at the bottom, and I thought it best that they should be first told this fact. My advice is to begin with a few plants, not too many. Do what you do thoroughly; this is the secret of success. As you gain in experience and become familiar with your plants you can branch out then. A dozen good plants well cared for are a delight, while a whole house full of starved, neglected things are a misery.

Now suppose you have one window on the south or east side of the house (preferably the south) which you can devote to plants. In it you can have about a dozen. I would not have more, for they would be too crowded. Flowering kinds are what you want for the most part, of course. Well, if I were allowed to choose for you, I would first choose the geranium, because it will stand more neglect than any other plant and do reasonably well; but it does so much better with good care that it ought not to be neglected for all that. Again, it is the least subject to the attack of insects, and the list of varieties is so large that every one can satisfy their taste. There is every shade of red, crimson, scarlet, salmon, pink and white. A few of the best varieties for Winter bloom are Master Christine, dwarf pink, a very fine bloomer; Mrs. E. G. Hill is a lovely blush, overlaid with a delicate lavender shade, double, and Mrs. Moore, which is pure white, with a ring of bright salmon around a small white eye. I would also have a rose variety, for its sweet-scented leaves are excellent for small bouquets.

The ivy-leaved geranium is a new form recently introduced, and has certainly proven to be one of the most glorious gems of the floral world, having glossy ivy-like leaves of a graceful and trailing habit, and the gay-colored flowers combined form an object of a most striking contrast and beauty. They are always robust and healthy and are of the easiest culture either in pots, hanging-baskets or open ground.

Abutilon or flowering maple is another desirable house plant. The foliage is clean and beautifully marked, bearing a strong resemblance to the leaf of our sugar maple. In fact, the whole plant looks much like a dwarf maple tree. It is a constant bloomer, and the flowers, which are bell-shaped, are of a drooping habit. The colors are white, red, pink, yellow, purple and scarlet. Like the geranium, this plant is not subject to the attack of insects.

The calla lily is too well known to need description, however it is shamefully neglected in treatment. Heliotrope is a favorite flower, and makes a handsome plant for the window, if it is well cared for. Its clusters of lavender and purple flowers are not only beautiful but very fragrant. It requires plenty of sunshine and a good deal of water, especially upon its foliage, which should be syringed or sprinkled daily.

Cyclamen is particularly adapted for window culture, and will give abundance of flowers with less care than almost any other plant. The colors are usually white, tipped at the base with rosy purple. A small pot should be used, and the crown of the bulb should be placed just above the surface of the soil, and set away in a cool place until the leaves are well grown. When the flower buds begin to show well, remove to a sunny spot, where they will open. This plant should be kept as near the glass as possible.

For many years the petunia has been one of the leading flowers in the garden, and few plants can give more general satisfaction for the window. The double variety is generally selected for window culture.

Carnations are the most magnificent of all the dianthus family. A rival to the rose. They are beautiful, fragrant, and give plenty of flowers a long time.

Oxalis is a good plant for hanging baskets; but to succeed

best with it for Winter bloom, the bulbs should be potted as early as possible in the Fall. The principal cause of failure of hanging baskets is that they are very apt to be neglected, and suffer for want of water. They are so exposed on all sides to the heated atmosphere of the room that they dry out very soon, and the water which is given them, quite often without taking the basket down, does not penetrate to the roots, but simply moistens the surface soil. The best and most satisfactory way of watering hanging plants is to take them down and set them in a pail of water for a while, when thoroughly wet place them in a sink to drain. Now a word in regard to soil. While it is true that many plants require a soil especially adapted to their peculiarities, yet a preparation made up of one-third leaf mold, if obtainable, and one-third turfy matter from under sods in old pastures, and the other third equal parts of sharp or builders' sand and ordinary garden mold. When I cannot get leaf mold, I use two-thirds more turfy matter and one-third made up of well-rotted manure and sharp sand. This mixture will need but little variation from the original preparation to suit most plants, which it is advisable to attempt to cultivate in the house.

For ivies, which are similar to roses in this respect, I would leave out about half of the turfy matter, and put in garden mold, as they require a more stiff and compact mold or soil.

Among the list of Winter plants the begonia should not be overlooked. Its cleanliness, beauty of foliage, combined with graceful flowers, and free-blooming qualities, make a most desirable plant. The begonia family is divided into three classes, and are known as tuberous rooted, flowering, and rex or ornamental-leaved varieties.

The tuberous rooted variety is very showy, and blooms profusely during Summer for a long time. During Winter they may be allowed to remain dormant, and about March can be repotted in fresh, light soil, and started into growth again. Amateurs who have but little room for the accommodation of house plants would do well to select this variety, because after blooming in the Summer and Fall it can be allowed to rest, by withholding water and put in some cool place, free from frosts, where it will remain till Spring.

The rex varieties, of which there are a dozen or more, varying in color and markings, are very effective as pot plants, and can be kept in perfection many years by cutting back when the joint stems become naked, making new plants of the cuttings, and repotting the old roots in fresh soil and allowing them to rest for a few weeks. Rex begonias are very partial to a warm atmosphere, and do well in the heated air of the living-room. They will drop their leaves if too cold or wet.

Flowering varieties make beautiful pot-plants for either Summer or Winter decorations, and a poor choice may not be made because all are exceedingly beautiful. Hybrida multiflora has flowers of a rosy pink color, and is a charming, graceful bloomer.

Begonias like a rather light soil, and a mixture of about one-third loam, one-third half mold, and the other third made up of well-rotted manure and sharp sand will suit them. Bad drainage and over-watering is the worst thing that can happen to begonias, or in fact to any plant.

Few house plants are more popular and give better satisfaction if properly grown than the Chinese primrose (*primula sinensis*). The flowers are all varieties of form and color, from the pretty single, so like the wild wood blossoms, to the charming double kinds, which resemble roses. Primroses delight in a cool place, with plenty of light and not very much sunshine, although they will bear a little if not too intense. A north window suits them best.

About the time the flower buds appear the plants should be set where they are wanted when in bloom, because they prefer to have their place assigned them and be undisturbed; and in watering care should be exercised that no water fall on the crown or cluster of buds, as it produces rot, and even the foliage is impatient of it if used too freely.

Changing them to positions of considerable difference in temperature should be avoided, for while the plants will endure 35° at night and 80° during the day, yet they become adapted to one place and do better if allowed to remain there. Do not crowd them among other plants, because they will not thrive so well as when given plenty of room.

The soil for primroses should be fine, light and rich; leaf mold, muck or garden mold which is found beneath the sods of old pastures, mixed with sufficient sand to make it light and porous, and a liberal supply of well-rotted stable manure. This will make a good soil for them. Where leaf mold cannot be had, chip dirt from around old wood piles will do. I would prepare the soil early, and when wanted for use thoroughly pulverize and sift it, because the rootlets of these plants are so very fine and tender that they cannot penetrate hard lumps of earth.

Young plants should be set in small pots. Using the soil as prescribed, pulverized and sifted; shade and attend to watering until well established in the little pots. Watering by immersing the pots to the rim in a pail of tepid water will be found to be the very best and safest way, for spraying them, even if very fine, will quite often break or damage their leaves, which are very tender.

As soon as the plants fill the pots with roots they should be repotted, changing to pots only one or two sizes larger, and treated as when in the smaller pots. Care should be taken not to allow the plants to become root-bound, because they will receive a severe shock from which it takes a long time for it to recover. Aim to keep the plants growing healthfully, and no more, if you would have flowers of large size and great beauty.

This plant seldom requires larger pots than four or five-inch, but should the roots become too much packed, give them a still larger pot, or else they will lose their vigor and produce small flowers.

To have the largest and best bloom, young plants should be raised each year. With a little care there is no reason why flower lovers should not be rewarded with profuse and lovely bloom all through the long, cold, cheerless Winter.

Some Annuals.

Perhaps some of our flower friends would like to cultivate a few flowers during the coming season, but are unable to devote much time to flower gardening on account of other duties. Under these circumstances I would cultivate a few annuals, for they will usually give the best satisfaction for the amount of money invested, and they bloom so profusely that the garden can be made bright and cheerful during the Summer and Fall.

If I were asked to select some of the best annuals, my choice would be as follows: Pansies, phlox and petunias. These three make very rich beds of showy, fragrant flowers, giving an abundance of bloom from midwinter until frost. The portulacca, which delights in a warm sun and sandy soil, are very effective when used for borders of beds. They make a very brilliant and gorgeous display when sown in masses. Nothing can be better for bouquets than sweet peas. They have delicate and fragrant flowers, varying in color from the brightest pink, and including the darkest purple imaginable. Th balsam, like the aster, is one of the most beautiful of our annuals—producing flowers of many colors and markings. Stocks, with their variety of fine colors, and large spikes of beautiful sweet-scented flowers are certainly very desirable for either garden or pot culture.

From this list quite enough could be selected to fill a moderate-sized garden, and will occupy all the spare time that can be devoted to them, if a good orderly garden is desired, and it should not be otherwise. Aim to have quality, not quantity.

Two Good Vines.

One of the best vines for the window or conservatory is the passion flower, generally catalogued as the Cassiflora. It is a beautiful, rapid growing vine, and is desirable for training about windows or a trellis. The flowers are curiously crimped, crape-like, and fringed, and are the admiration of all who see them. They are exquisite in color, being a rich blue, marked with brown or green, and are very delicate in texture. The leaves, which are of a bright green, grow similar in shape to the American ivy, and are five parted, but it will be seen by the accompanying illustration that the edges are smooth, while the leaves of the latter plant are notched. The vine would be well worth cultivating for its foliage alone, but when laden with flowers its beauty can be better appreciated than described.

The bovardia is a beautiful plant when well grown, and is deserving of much attention. The plants are quite easily grown, and will reward the grower with a profusion of beautiful flowers. They are of a shrubby growth, with flowers borne in clusters. To have plants in the proper condition for Winter blooming, it is best to pot young plants in the Spring and encourage a free growth by giving a weekly watering of liquid manure and never allowing them to become root-bound before changing to a larger pot. During the Summer's growth the back branches should be cut occasionally to make the plants bushy and compact, which will also produce a larger flowering surface. Pinch all flower buds that appear during Summer. When the time comes to house plants in the Fall, give this one a warm place in the window and shower its foliage frequently to keep down the red spider. Davidsoni, white, leiabtha scarlet and rosea multiflora, pink, are the best varieties.

Among the flowering vines there is nothing so desirable for sitting-room culture as the hoya, perhaps more generally known as wax plant. It has large, pointed ovate leaves of a very thick texture, which enables it to stand dust, dry air and heat, quite as well as the ficus or India rubber plant. The flowers, which are of a flesh color with a darker centre, are star shaped and borne in clusters. They have a delightful perfume, and hang on the plant a long time before fading. A peculiarity of this plant is that the flowers are produced from the same peduncles or stem of the flower cultured each year, and after the blossoms have fallen, the little stub that is left to mark the place where they were born should never be removed, as would be supposed, because it will destroy future crops of flowers. Should you want to give flowers of a plant to a friend, cut off single flowers, but not a cluster, on the main stem.

In its growing season the hoya should have the highest place in the window, so as to get all the heat possible, but should be in the shade of another plant, so as not to be exposed to the full and direct rays of the sun. Spraying the leaves frequently will cultivate much of their luxuriance, and will assist in keeping down the mealy bug, its particular insect enemy. Let the soil be rich, sandy and light, and do not over water, but give it just enough to keep the soil nice and moist all through.

It takes this plant some time to get fully established, and sometimes, after it is rooted, it will appear to stand still for a while, and will make one impatient, when all at once it will begin to grow, and will sooner or later bloom.

This vine should have a stout trellis, because the leaves are quite heavy, and usually the ordinary sort are too slender to afford the plant a proper support. In season of rest, it will bear to be kept rather dry, and moderately cool.

Watering Plants.

How often do you water your plants? This question has been repeatedly put to me, and while it may seem to be of no great importance to the eyes of some flowergrowers, yet it is a very important part of floral culture.

Many persons mechanically soak their plants, just as if they were a piece of machinery, which requires oiling so often, whether dry or not, and others fall into habit of giving a little at a time, and often. The first method is not so bad as the last, if good drainage has been provided, which should always be the case, because all surplus water will naturally run off through the hole in the bottom of the pot, the soil only retaining sufficient water for the plant; but where a little at a time is given, even if it were often, thus they are led to believe it is wet all through, but if the soil should be turned out of the pot, they would probably find the lower half as dry as dust.

There is only one safe rule to go by in watering plants, and that is this: Whenever the surface of the soil appears dry, then give sufficient water to thoroughly wet the earth all through.

I find the best and most satisfactory plan in watering plants is to place them in a sink, there they can be watered to advantage, and without wetting the floor. It is always a good plan to do the watering in the morning before the sun shines on them, or in the evening after the sun has gone down. I believe one reason why plants do not give satisfaction in the sitting room a great many times is, because the leaves are allowed to become very dusty from sweeping. The leaves of plants are the lungs or breathing apparatus, so if the pores of the leaves are allowed to become stopped up with dust, what is the result? It would have a similar effect upon plants, as it would upon humanity if their lungs become clogged, producing a sort of drying or wasting away of that most important organ, and a consequent gradual decline in health. So if we are to have vigorous looking plants, we should keep their foliage clean, which can be done by syringing them or washing them. A greater proportion of the dust can be prevented from settling on the plants while sweeping if a light cloth or newspaper be spread over them carefully. A word to the young people. Among my readers I presume there are a good many young people. I wonder how many there are who like flowers, or are interested in their culture. Perhaps not many, but I am quite sure that if all knew how much pleasure there is in cultivating a few flowers, no home would be without its flower garden, provided of course their circumstances would allow it. There is a great deal of valuable knowledge to be acquired from the study of plant growth, and they certainly have a refining influence upon the mind and character of those who become interested in their development day by day.

Some people are under the impression that there is a knack

about growing flowers, and unless they possess a peculiar gift they cannot succeed with them. If you really love flowers enough to give them such care as they require, you can grow them easily. Grow a few flowers the coming season and observe the results yourself, but by all means do not plant too much to begin with, as is the case with many beginners. It it far better to begin with a few less particular plants before undertaking the care of some which require more careful treatment. As you gain in experience and become familiar with the requirements of the plants under your care, it will do to branch out.

Herbaceous Plants.

Flower lovers ought not to forget to give some attention to herbaceous and perennial plants and their cultivation. This extremely useful class of flowers will afford a great deal of pleasure to the persons who plant them, at a small outlay of labor; and after they are once planted they require but little further care for a long time. When they have become well established they are pretty sure to give an annual bloom.

Herbaceous plants, like many others, have both their advantages and their disadvantages. There is little weeding to be done that cannot be accomplished with the hoe; no seed beds to make in the Spring, and sowing seeds and the work necessary to be done among them each season will not amount to as much as that required by a bed of annuals. So far, it would seem that the argument is in favor of herbaceous plants, but when you consider the fact that annuals, as a rule, give a supply of bloom during the whole Summer, the argument seems to be strongest on the other side. To those who have unlimited time I would advise the cultivation of both.

To those of my readers to whom the time is limited, it would perhaps suit their circumstances best if they were to grow the herbaceous kinds, among which there are many very beautiful and desirable kinds.

Herbaceous, or border, plants require rich, mellow soil in which to grow, and the weeds and grass should be kept from choking them.

Work in an annual application of well-decayed manure around the root of the plants. See that these hints are attended to, and they will flourish and increase in size and beauty for a long time. Chip dirt from around old wood piles is excellent to put about these plants as a mulch, and it is usually coarse enough to be of value in keeping down the grass and weeds.

Perhaps some of the plants here described do not come under the title "Herbaceous Plants," if spoken of with botanical accuracy, but I will classify them all together as a matter of convenience.

Decentra, or perhaps better known as bleeding heart, is a very graceful plant. It has very fine foliage and flowers, which are borne in racemes about a foot long, and droop in a very graceful way. Flowers are heart-shaped, being pink with a creamy white centre.

What is more excellent than the daisy for the border or low beds? It is a beautiful little plant that produces very double white and pink flowers. The florist's daisy is not the same as those which grow wild in our pastures.

The pæony, with its massive fragrance and grand blossoms, is a very satisfactory and hardy plant. What a trio of excellence. This flower, like a few others, seems to carry one's thoughts back to old times and associations. It ought to rank among the first of our hardy plants.

There are many more varieties now than when our grandmothers grew them. The old red, which society has named P. Ruba, P. Rosea, and P. Officinal, are good varieties. I imagine you will say, "Oh, they are as old as the hills; I want new kinds." Well, please don't discard these old friends to make room for new ones, without a little hesitation. If you will have new kinds, then grow both. Among our new varieties are P. Fragrans, one of the best pinks; Strata Speciosa, white centre, pale rose, very sweet; Pattsii, purplish crimson, and festive white spotted with carmine.

Pæonies are not rapidly growing plants, and until they are well established do not show what they are capable of doing. They may be planted either in the Spring or in the Fall, and when once set do not meddle with the roots, because they do not like their roots disturbed. After making a good growth, which takes at least two years, they will reward the owner with rich colored flowers.

There are few more desirable plants for a bed on the lawn than the hollyhock. Many have discarded this old valuable plant to make room for something new, because it is not the fashion. However, I feel that the time will come, if not already on the way, when many of our old time-honored flowers will rank among the first in our flower gardens.

The tall growing varieties of the hollyhocks are desirable for backgrounds, and are very effective when planted along fences, or in front of old wood piles, which it is desirable to hide, and when used for this purpose, the double kinds, which are dwarf growers, can be used in front of the tall ones to good advantage, and thus a perfect bank of bloom can be had.

Japan anemone is a fine plant for late blooming, coming in nearly after everything else has gone. The best effect can be had by growing them in clumps about the roots of trees.

Give them a few years to establish themselves and they will surprise you some Autumn day with a delightful bloom, when most flowers have faded.

Astilbe japonica (spirea) is a hardy herbaceous plant which bears delicious trusses of feathery flowers, and is handsome as a border plant. Florists use it for Winter fencing extensively.

Delphinium (perennial larkspur) is a good blooming plant, bearing flowers of many varieties, and shades of colors. The foliage is clean and pretty, and the habit of growth strong and good. Formosum is an old variety with blue flowers of exceeding richness. There are newer varieties which in tone are lighter.

A good clump of delphinium will often measure six feet around, and each stalk for a third of its length will usually be covered with blossoms. The Sweet William is a very old, and once popular flower, but during the rage for bedding plants has been somewhat forgotten However, its merits will again be appreciated when people get tired of bedding plants and it will rank among the best of our old-fashioned flowers, when they again become popular in the flower garden. There is nothing more hardy or more sure to grow and grow well. It has been greatly improved upon in the past few years, the blossoms now being much larger than of old, and of exquisite color and markings and some are very double like miniature roses.

The odonis is a good plant, bearing a bright yellow flower, and having finely cut foliage.

Rocket is a very desirable plant. It needs no description, for it is too well known.

The perennial phlox produces a very brilliant show of flowers. It is perfectly hardy and the flowers are borne in large masses from two to three feet in height.

Lilla violet, Adelina Patti, white eye, rose color, Dutchess of Sutherland, pure white, are fine varieties. The yuccas are an interesting class of plants. They give to the garden something of an oriental or tropical appearance that is exceedingly pleasant. They will survive most of our Winters, if well covered with litter or leaves in the Fall. There are several varieties of this plant, but Y. filamentosa is the hardiest. It will send up a strong flower stem during the Summer, bearing a large spike of whitest flowers. Yuccas are evergreen perennials, and they delight in a rich soil.

There are other good herbaceous plants which should be named, but it would take up too much space to describe all of them here. From the above list enough of them can be selected to fill all the space in an ordinary sized garden that you would care to devote to this class of plants.

Most of the plants above named are hardy enough to stand our coldest weather without protection, but they will do enough better with protection to pay for all the trouble in giving it. If not protected the continual thawing and freezing exhausts their vitality to such a degree, that it takes them most of the following season to recover from the bad effect. Give them a light covering in the Fall, of litter or evergreen boughs, and in the Spring, when danger of severe frosts has passed, remove the covering and your flowers will come out strong and healthy.

Ornamental Climbers.

One of the most interesting and useful classes of garden plants are the climbers. For making screens, and covering fences, arbors and verandas. In the flower garden, and the adornment of rural homes, no class of flowers are more useful than these, and no drapery devised by the highest art of man ever equaled the glorious drapery with which nature beautfies the stump, the ruin or the tree.

Among the climbing annuals there are two which ought to be in every collection—the morning glory (*convolvulus major*), and sweet pea. The former is well known, and the only fault to be found with it is, that the flowers are open only in the early part of the day, being brightest at about sunrise, and in order to enjoy their beauty one must arise early. However, the sight of a good bed of these flowers on a dewy morning, with the bright sunbeams shining full in their faces, is quite enough to tempt any lover of the beautiful to arise early and enjoy their glory. The sweet pea is adapted for training over fences or low trellises, and for large bouquets nothing can be better, because the flowers are lively, delicate and deliciously fragrant. It is a profuse bloomer. The cobola scandens is another good tender climbing plant. It has fine foliage and bears large bell-shaped flowers, which are green at first, changing to a deep violet hue. Its growth is quite rapid, and if strong plants are set out early in the Spring and in good rich soil, they often grow twenty feet in length and branch freely.

The only objection to this plant is that the seed does not germinate as readily as that of other plants. It requires care and favorable conditions, and no great amount of success need be anticipated by sowing seed in the open ground. Start them in the house or hot bed by placing the seed in moist earth, edge downward, and do not water until the young plants appear, except that the earth becomes exceedingly dry. If care is taken, and good judgment exercised, there is no reason why this plant cannot be easily grown from seed. Plants may be removed to the house in the Fall, if desired, where they will do reasonably well. Tropæolum, or nasturtiums, are among the very cleanest, prettiest and best climbers. The flowers are so plentiful, the colors so varied, and the comparative ease with which they can be raised, are qualities which speak much in their favor. T. Lobbranum is a very rapid grower, and bears brilliant and colored flowers. Caroline Smith, spotted; Colour de Bismarck is a peculiar shade of brown; Napoleon III, yellow and red. There are also dwarf varieties, which grow about a foot in height, that make very attractive beds when grown en masse, and are also excellent for rockeries. Crystal palace gem, yellow spotted with maroon, and King Theodore, very dark, are two good varieties. Nasturtiums are very desirable to work into bouquets, and the peculiar fragrance, which is delicate but pronounced, make them a special favorite. A soil that is rather stiff but not too rich will suit them best.

The Madeira vine is a tender tuberous plant with thick, glossy, light green, almost transparent leaves, having a root or tuber very much like a potato. It is a splendid climber and will grow to a remarkable height during a season. The tubers should be planted out in the Spring, when they will grow at once, and, if in a warm, sheltered place, very rapidly, until the slender branches with their pretty leaves have covered a large surface; and when I add that it bears white flowers, which are quite delicate and very fragrant, I will leave you to judge whether it is a desirable plant to grow. In the Fall the Madeira tubers can be taken up, first removing the tops, and stored in a cool but frost-proof cellar.

Ampelopsis quinquefolia (Virginia creeper) is in my estimation equal, if not superior, to the well-known English ivy. It is perfectly hardy and ornamental, easily transplanted, a vigorous grower, and one that will flourish anywhere, on any soil, and almost under any condition. It furnishes the most dense and graceful shade of any plant, and is quite free from insects. It is supplied with many spiral tendrils, which are almost as strong as wire, and the firmness with which they grasp any object makes it quite able to support itself well in almost any situation, and to defy the fiercest winds and storms. In the Fall the leaves change in color to a bright crimson hue, which for a short time is very effective, and before falling are of the deepest scarlet.

The wisteria is another very beautiful climber, bearing hundreds of long, pendulous racemes of delicate light blue flowers, which often measure twelve inches in length. The flowers appear about the last of May, and before the leaves; at least, before the leaves have made but little growth. The wisteria is said to be hardy and able to withstand our Winters, but I prefer to put a light covering around the roots for protection.

Window gardens and their arrangement require thought and care, as well as any other feature of domestic economy. I am often asked the question, "How do you manage to have such beautiful plants?" and I often feel inclined to answer, as did the doctor in the case of the lady who had been taking too much medicine, that all that was needed was light, water and air. These three words involve so much meaning that perhaps it would be well to give a more definite explanation. Well, no exact rule can be given, but the first thing is to see that the plants are kept moist, not only on the surface of the soil, but all through. The leaves should be sprinkled at least once in a week in order to keep the pores open and free from dust. On bright days, open the window and let in the fresh air, but the air or draught should not blow directly on the plants. A thermometer should be kept in the room which ought to register about 60° or 70° during the day and about 45° at night. Admit all the light possible, and place all soft-wooded plants nearest the window. They must have light or they will lose color and vitality and grow crooked and one-sided by reaching toward the light. Many plants, such as primrose, narcissus, hyacinth, and other bulbs, will do better in a room without fire, provided the atmosphere is above freezing. Ivies, hoyas and passifloras make good climbers. Ivy geraniums, smilax and oxalis will do well as basket plants, and geraniums and begonias of the flowering variety will bloom to the best satisfaction.

For the sunless window, flowering plants are not likely to be successful, but there are other plants sufficiently attractive, without the aid of flowers, that will thrive well in such a window, and for this purpose the India rubber tree is a good plant. It grows slower in the house than if kept in the greenhouse, but after a few years' growth you will find that it will take up considerable room, even in the sitting room.

Ardisia and aspedistra are also good plants where strong, bold foliage is desired. A shady window offers a fine chance to grow a few ferns, but they must be protected from the heated atmosphere of our rooms by means of a fern case.

Nearly everyone who has grown flowers has learned something about them which they have not read about, and I would be glad to have you send in hints and suggestions drawn from personal experience, that would be of interest to your sister flower growers.

There are some hints that may be properly suggested about flowers and their arrangements, but as for instructions, none can be given with an assurity that, if followed, they will always lead to success. It requires good taste and some knowledge of the harmony of colors, and these cannot be purchased or taught, although they may be cultivated and developed to a certain extent so as to produce good results. Some persons have the knack of arranging flowers tastefully and naturally, because they are gifted and work from a natural basis, while others cannot excel, for the simple reason that they do not possess the faculty the successful person does.

It is a great thing to discover and bring out the ornamental side of what is called homely and common—weeds if you like—a gift which only some favored few possess; but those who have it can clothe their surroundings with beauty in spite of the most unfavorable circumstances. Thus dandelions are not generally valued, except for the amusement of children, who love to fill their aprons with the " pretty stars in the grass," but take an ordinary stone butter jar, the deeper and darker blue the better, avoid the abomination of decorating it in any way, and into this jar drop a mass of the golden blossoms, and then you will have a bit of color that every artistic eye will appreciate.

The main idea in arranging flowers is naturalness. Avoid all artificiality and everything that will take the attention from the flowers themselves, and aim to make them look as much like growing on their own stems as possible. Another point is not to put too many flowers in one vase, or torture them into shapes and positions that Nature never put them into. It is too common a habit to crowd a mass of blossoms, with very little foliage to relieve them, into one-third of the space which they would naturally occupy.

By way of illustration, which will perhaps give you a better idea of what I mean, suppose you cut a branch of sweet peas with long stems, not too many for they must not be crowded, and do not attempt to arrange them, for it is a characteristic of this flower that it never can be anything but graceful under any circumstances; you want just enough to fill your vase and allow them to bend about naturally, which they cannot do if crowded in the least. Now drop them into a vase with a flaing mouth and they will arrange themselves in such a way as to delight the eye of an artist; some will droop, others will remain upright, but the general effect will be airy, graceful and delicate. If you were to add to this beautiful vase of flowers a cluster of scarlet geraniums or roses, where would be the delicate effect? Gone. There is no harmony between the two flowers, and the addition has taken away the beautiful effect produced by the sweet peas alone. The roses and geraniums when taken by themselves are beautiful, but when combined with sweet peas the good qualities or effects of both are lost.

Take this same cluster of roses and put them in a bowl with nothing but their own leaves and you will have another beautiful bouquet that will be appreciated by all who see them. If you should add a stalk of gladiolus the effect of the roses alone would be spoiled. The gladiolus, like roses and geraniums, are beautiful flowers, but they do not harmonize well together.

Again, scarlet geraniums and salmon ones do not combine well, but any white flower can be used with scarlet, and deep-toned orange or brown flowers, like coreopsis, can be effectively grouped with the salmon varieties.

When it is possible, flowers should be surrounded by their own foliage, and the rich green of perfect rose leaves is particularly handsome. Some writers say that roses should never be put into vases with any other flower. This may be applied as a general rule, but there are exceptions to most rules, and to this I will make one exception. Say you had a vase of pink roses, and desire something to increase or bring out the good effects. Add a flower of the wild clematis or virgin's bower and note the effect. The clematis, with its delicate, airy nature, gives precisely the unstudied grace which any vase of flowers should have.

If you practice on such combinations and study their effects carefully, you will understand better what I mean. In arranging flowers always bear in mind that the dish or vase chosen to contain them should be of a character to suit the flowers, as, for instance, short-stemmed flowers like balsams show to the best advantage in shallow dishes, while tall flowers should be given tall vases.

There can be no greater sign of human progress than the evidence all around us, of the earnest, energetic people becoming imbued with the love of plant life. Flowers are emblems of refinement, purity and happiness, and I have observed that those families who cultivate them live in peace and love. Quarrels and continuous strifes, which so often curse and blight domestic life, are seldom known where flowers are grown. Does not the experience of mankind largely, if not fully, confirm the truth of this observation? They possess a charm that seems to temper the hardness and sordidness of earthly life, and excite in the mind the tenderest, most kindly, innocent and cheerful thoughts. The pleasure of cultivating a flower garden, and the particular love for its products, afford a quiet observation and thought, and there is pleasure in all the work. One must give care and companionship to plants and bowers in order to learn their graceful nature and feel their beneficial influences. They are unrivaled, too, as friends and companions, and their conversation is always in charity and good sense.

The importance of this subject, seems to me, would justify a careful examination into it, and if the result of that examination be given to the world, I feel sure it would be favorable to the most beneficial influence of flowers on domestic happiness.

Every family can surely find a little space for flowers, and if it is only a few feet square you can have many choice plants.

The labor and care required to keep a few flowers in proper order is not as much as people imagine, and their missions, like the angels, are pure, while their color, symmetry and fragrance attract the eye, delight the sense of smell and kindle a taste for the beautiful.

A pot or box of flowers in a city window contributes to the happiness of all who see it, and the fragrance which ascends from their swaying blossoms has a tendency to sweeten not only the atmosphere of the room in which they may be, but the very nature of the occupants of the house, though it should happen to be a humble tenement.

If you have plenty of flowers, give them with a liberal hand to friends and acquaintances who have none, and, above all, to the sick. Perhaps a few of our readers can appreciate the pleasure and encouragement a gift of flowers must be to those who are unable to be out of doors and enjoy the beauties of Nature. A gift of flowers can be seldom unappropriate, if ever, either to young or old, and purity and goodness are painted on every petal.

There has been a great deal written about the care of plants in the sitting room, and how they should be treated so as to be a pleasure at all times, and yet the subject will bear further comment. I think one reason for dissatisfaction is due to the fact that many times they are allowed to become too dusty from sweeping. This may be partially avoided if a light cloth or newspaper is thrown over at that time, and occasionally the leaves may be washed in clear water which will leave them clean. Then, too, the art of watering plants requires careful study, for some plants need wetting much oftener than others, and it is seldom that a whole collection requires water at the same time.

The amateur, however, who really loves flowers, will readily learn this, but there is one thing I wish to call particular attention to, and which is of great importance in plant culture. It is to provide perfect drainage, that is, place an inch of broken crockery or small pebbles in the bottom of the pot, so that all surplus water will run off. If plants are given what they want or need, but above all, fresh air, I see no reason why you should not succeed with them.

A word in season. It will soon be time to make arrangement for the Fall planting of bulbs, and before giving a list of the most desirable sorts, I wish to make a few preliminary suggestions. October is considered the best month for out-door planting of bulbs, although they can be set any time later, so long as you can get good bulbs and the frost will allow you to put them in the ground. It is not, however, best to wait for the last chance. The soil in which to plant bulbs should be light and sandy, and enriched with plenty of well-rotted manure, which is the best of all for bulbs, and may be used in any condition. It is a good idea to raise the bed, in which bulbs are planted, a few inches above the level of the yard, and if a little higher in the middle than at the sides it will be an advantage.

The potting of bulbs for blooming in the house may also be continued through the month. A good soil for this purpose, and one which I have used with success, is composed of one part leaf mold from the woods or fence corners where leaves have drifted for a long time and rotted; to this add one part sand and two parts soil from beneath the sods of old pastures, and mix the whole thoroughly with a small quantity of well-decayed manure. This compound requires but little variation of the ingredients to make it a proper soil for the cultivation of most of our flowers, and I would advise all who intend to grow flowers to get a little pile on hand. You can keep each ingredient in a separate box or pile, if desired, and mix as wanted for use.

After bulbs have been potted, they should be set in a cool, dark place for a few weeks to form roots; so that when you bring them to the light and warmth, they may receive nourishment from the soil to support the growth of leaves and flowers. If you want something to brighten up a window and make your house cheerful and home-like, don't fail to have a few bulbs. Let each of my readers try half a dozen, at least, whether he or she be interested in flowers or not. Encourage your children to love flowers and grow them. You will make better boys and girls of them, and the small amount invested in flowers will never be regretted.

Bulbs for In-doors and Out.

August, the last of the Summer months, is gone. Another season of buds and blossoms will soon be numbered among the past. The ripening leaves are about to put on their gala dress of gold and scarlet before bidding us a long farewell, and the Autumn storms, and Winter cold, will soon compel us to retire to the sitting room and parlor. Here we may seek pleasure in the society of books, plants and friends.

I have carefully prepared the following list of bulbs for Fall planting in the garden, and for forcing in the house. The description of each is brief, which must necessarily be, in order to mention all of them in the space devoted to this department. They are all charming flowers and excellent for decorating purposes to brighten up the window in Winter and make home cheerful. All are beautiful and desirable, but, if there is one that I admire more than the others, it is the Bermuda or Easter Lily. One of these bulbs should, by all means, be in every collection.

The Bermuda or Easter Lily.

This is a favorite flower everywhere, and one that will give satisfaction. It was exhibited at N. Y. Horticultural Society in 1881, by Prof. W. K. Harris of Philadelphia, Pa., and was named Lillium Harrisii. Since then it has been extensively used for decorative purposes at Easter, and the general public have substituted the name "Easter Lily." By experiments it has been found that bulbs of a superior quality can be produced in the Bermuda Island, than when raised in our dryer climate, and so remarkably does it succeed there, that it is now known as the Bermuda Easter Lily. It is a very free bloomer, and is not difficult to succeed with, as some people imagine. Give it the same soil as recommended for hyacinths and tulips, and after potting set in a cellar, or some place where it is cool and dark, until roots have formed. When you bring it to the window, give it the coolest place. Do not allow the direct sunlight to strike the plant, and if you have provided good drainage, you may be reasonably sure of success. If you desire them by Easter, the pots should be brought up by the middle of February. The flowers are large trumpet shaped, of the purest white and delightfully fragrant. Can anything be more beautiful? I much prefer them to hyacinths or tulips, and would advise all those who love flowers to try one, and before the season of bloom has ended you will wish you had half a dozen instead of one. Sometimes the lily is attacked by green lice. Should you discover any, take particular care to keep them from damaging the plant, which can be easily done, by washing or syringing the leaves and stalks with a decoction of tobacco. Avoid the use of this after the flowers appear, because it will stain the petals.

The Freesia.

This bulbous plant is a native of the Cape of Good Hope, and was introduced into England many years ago, but for some reason it dropped out of cultivation, and little has been known of it until recently. For Winter blooming it is one of the best. The flowers are yellow, sometimes pale or even creamy, with an orange blotch on each of the lower divisions of the perianth. Some are white, occasionally showing a few violet lines on the lower divisions of the flower. They are delightfully fragrant and each stalk will bear a cluster of from two to a dozen flowers, which last for several days. They are very easily grown in pots the same as tulips or hyacinths, but instead of setting one bulb in a pot, you can plant half a dozen in a five inch pot, because they are so small. The foliage is not unlike that of the gladiolus, only much smaller. The Freesia will give much pleasure for the money invested, and it would be well to have half a dozen pots. After they have completed their growth, which can be told by the dying leaves, take the bulbs from the soil, wrap them in paper, and keep in a dry, cool place during the Summer. In the Fall you can repot them again.

The Hyacinth.

Among the Dutch bulbs the Hyacinth is perhaps the most beautiful, fragrant and most popular. No flower has done so much to make cheerful the tedious Winters of northern countries as this. When a plant is in full bloom in the house it will send forth its delicious perfume and fill the air of a room with breath of Spring. Hyacinths differ in habit very much. Some varieties throw up strong flower stalks, with a bold or rather loose truss, while others have but a short stem with a compact truss. My preference are the single varieties, because they have larger spikes and the flowers are not so crowded together. This is, however, a matter of taste, and there are others who may prefer the double varieties. There is a variety called the Roman Hyacinth, which is not as popular as the ordinary sort, but with those who love flowers for real beauty, it will become a favorite when once known. The flowers of this variety are somewhat smaller than the ordinary hyacinth, but what they lack in this respect is fully made up in gracefulness and quantity, sending up, as they do, from one to five flower stalks from the same bulb. The flowers are not so closely set along the stem, and are given a chance to show themselves, while the others grow such short stems that the flowers are not given chance to develop, and much of their beauty cannot be seen.

Narcissus.

This is one of the most popular bulbs for forcing. The flowers are very delicate and emit a delightful perfume. Perhaps some of our readers are familiar with the fabulous story connected with this flower, but for the benefit of those who may not have heard of it, I would say that it runs to the effect that Narcissus was the son of Cephisus, one of the Grecian River gods. He was uncommonly beautiful and fell violently in love with himself on seeing his figure reflected in the fountain, and wasted away with desire until he was changed into the flower that bears his name.

The varieties of the single and double Narcissus are hardy with us at the North, but the polanthus family are tender in this locality in the open ground and should be kept for Winter forcing. For forcing, one bulb is sufficient in a five or six inch pot. There is a variety of Narcissus of the polanthus family that is comparatively new to many flower lovers; it is grown extensively in China as a national flower, for blooming at Christmas and New Year, and is known as the Sacred Lily. The bulbs are very large, and each one throws up many flower spikes bearing clusters of pure white flowers, with a yellowish centre, which are quite fragrant. They can be grown in soil or water. The Chinese usually flower them in a bowl or some vessel, filled with pebbles, in which the bulb is set. The dish is then filled with water and set in a light sunny window, where it blossoms in two or three weeks. It is not always that one can get the true bulbs in this country.

Tulip.

The Tulip has been a favorite flower for many years, and it comes as a matter of course in such variety that it is an easy matter to suit all tastes in the selection of color. You could invest a good many dollars in named varieties without duplicating the bulb; but I would not advise you to do this, for some of the cheaper sorts are quite as satisfactory as the higher priced ones. Named varieties are sold at fancy prices because they are new, and not because of special beauty. Mixed collections can be had at very reasonable prices, and contain nothing that it is not well worth growing. The variety, known as the Duc Van Thol, does well in the house; but, as a general rule, it is best to leave the selection of variety with your dealer, unless you are experienced in this line, simply telling him whether you desire them for pot or garden culture, and what colors you prefer.

The Crocus and Snow Drop.

The Crocus is an interesting Spring blooming plant. It is of low growth and blossoms much earlier than any of the other bulbs. It is often used for borders along the edges of walks and beds, and it is one of the few plants that can be set in the grass to take care of itself year after year, and without the least fear of being crowded out of the grass. Few other garden plants give signs of awakening from Winter's slumbers when this comes into bloom, and only the Snow Drop is earlier. The Snow Drop may be treated like the Crocus, and one of the best uses that may be made of both is to set them here and there in the grass during the Fall. There are other bulbs that are desirable for Fall planting in the garden; but to describe each separately would take up too much space, so I will simply mention them and you can find the description in some catalogue. The Lily of the Valley is known by everyone, and a sweet little flower it is; but the Scilla and Crown Imperial are not so well known, yet they are worthy of notice and perfectly hardy.

The Grape Hyacinth is an interesting plant bearing little blue and white flowers. Then there is the Iris or Flowering Flag, as it is sometimes called, and the Anemone, which produce very pretty flowers, as all will admit who have seen them.

Decorations for the Holidays.

At the approach of the holiday season, the subject of room decoration is one of interest almost everywhere, but in many communities it is often puzzling about this time to decide as to what shall be used for the season near at hand. Flowers are usually scarce at such times and involve an expense too great, that it cannot be afforded in many instances and the question presents itself, "how shall we trim?"

Well, we must first have something to trim with, and the foundation of most of the work, which I propose to outline here, will necessarily contain evergreens of different varieties, such as Cedar, Spruce, Pine and Hemlock. These are all sombre colors, I fancy you will say, but by a little thoughtfulness, at the proper season, bright Autumn leaves and the berries of some of our wild flowers may be gathered from the woods which will relieve this sombre effect produced by the evergreens. Some people make a habit of gathering ferns and Autumn leaves to press, and if any of our readers, who are interested in the coming festive decorations, have been thoughtful enough to gather some the past season, here is an opportunity to use them; for nothing makes prettier or more fairy-like decorations than ferns, and the bright Autumn leaves work up elegantly. The large leaves of the English Ivy and Ivy Geranium, where such are at command, taken from the branches, can be used to form many pretty designs; or the growing vines, with their pots concealed, may be draped over pictures and windows in a very artistic manner. Then we have the bright fruits of the Wild Rose, Climbing Bitter Sweet, Sumac and Mountain Ash, while the different varieties of everlasting flowers furnish much valuable material. In some parts of the country grains or grasses, if gathered before they become too old and carefully dried in the shade, can be used to great effect.

A careful summary of the materials here mentioned will reveal the fact that we have a pretty good stock, and which costs nothing but a little forethought, and care in gathering them at the proper time. In decorating, festoons of evergreen can be made to bring about very satisfactory results if judiciously arranged. These are made with a stout rope, and fastened in place by winding twine or fine wire about them as fast as the evergreen is put in position. Care should be taken to wrap the twine firm, if you would have a substantial wreath. An occasional cluster of the bright berries or fruits, previously mentioned, may be worked in the festooning in such a manner as to produce very pleasing effects, but unless you have plenty of them I would advise using them sparingly, or perhaps a better way would be to wait until you have the festoons in place, and then add the little clusters of berries or everlasting flowers where they will stand out in bold relief.

Nothing is more suggestive or pretty than mottoes, and those who possess the art of making the letters for this purpose can render valuable assistance to the decorator. The letters should be cut from heavy card-board with a sharp knife, after first determining the heighth you want them, and marking out the letters with a lead pencil. The surface of each letter is next covered with small branches of evergreen, and fastened in place with a needle and thread or glue, and afterward trimmed into shape with the scissors. If some of the bright hued materials of whatever you have to work with is worked in the letters, it will give a bit of color that will contrast effectively with the more solemn tints of the evergreen, and, where judiciously used, cannot help but please the eye.

In most communities there are persons that can make real pretty paper flowers, and where real flowers are not obtainable, vases of these paper flowers mingled with dried leaves, grasses, or everlasting flowers, can be displayed in a very satisfactory manner, while groups of bright Autumn leaves interwoven with ferns may be securely sewed to heavy paper or card-board and fastened to the walls of the rooms here and there, to relieve the bareness. I would suggest that the card-board used for this purpose be as near the color of the wall as possible. This will give your work the appearance of being attached directly to the wall, instead of on paper.

Two very pretty designs for festive decorations are a star and a cross. The star can be made quite easily out of laths or similar pieces of wood, nailed together so as to form a triangle. Make another triangle of exactly the same size and shape, and nail both triangles together in such a manner as to bring the point of each opposite the centre of each lath that form the triangles. This will form a six-point star that will be strong and durable. This may be covered with evergreens in the same manner as the festooning and a small cluster or something bright may be fastened to the intersections of the triangles. A small bouquet can be suspended in the centre of the star by two fine wires crossing each other like an X, thus giving it the appearance of being finished. A very pretty star can be formed out of large ferns in the following manner: Select from your collection of pressed ferns six of the largest and most perfect and having the straightest midrib, make them all the same length, and after drawing a circle on a sheet of stiff paper or card-board about two or three inches in diameter, arrange them about the circle as uniform as possible, with the base or widest part of each toward the centre. The space in the centre may be filled with a small bouquet of artificial flowers that will combine effectively with the light green of the ferns. This will make a six-point star that will attract attention and admiration for its peculiarity.

A good cross is about as effective to make as any of the festive designs, but I will endeavor to outline one, so that with a little patience and some knowledge of the use of the saw and hammer you may construct a very reliable cross. The frame work does not of course require a very fine display of wood or workmanship, inasmuch as it is hidden from view, so you need not be discouraged if your work is not as smooth as that which a skilled workman could produce. Now, for an illustration, suppose we take four laths and nail them together in the form of a flat cross; that is, have two laths perpendicular and two horizontal, leaving a space between them of say two inches. Take another frame the same size and shape and then fasten small blocks or squares of wood between the two frames at the top, bottom and arms. You will next want a base. This can be made out of two boxes, one being a little larger than the other. Nail the smaller box to the upper side of the larger one, leaving an equal space on each side. To the centre of the small box fasten the frame, and the cross is ready to be decorated. Some strips of cloth can be wrapped about the frame work and tacked, to which small branches of evergreen may be firmly and smoothly sewed. This part of the work should be neatly and carefully done, so as not to appear rough and uneven, and if necessary it may afterwards be trimmed a little with the scissors. Add here and there a bit of something bright to give an expression, and, if obtainable, an ivy draped about it, with the pot concealed, will produce very pleasing results. For the base, nothing is better than moss and ferns. As a usual thing they can be found quite near to where the evergreens grow, and whoever is elected to go for the evergreens can at the same time procure the ferns and moss. The ferns should be kept in a cool cellar till wanted. Where live ferns cannot be obtained, pressed ones could be used instead, but not with quite as good results, of course.

A very pretty device for the wall is a floral pocket. This is usually constructed of willow, and resembles an oblong basket cut lengthwise, with the wicker work turned up on one side to form the back. Some florists, I believe, keep them on hand, but anyone for that matter, can easily make a pocket or rack out of card-board, similar to those used for holding newspapers, only smaller of course, that will be as serviceable as one made of willow. The outside of the pocket can be covered with evergreens, which will hide the material that it is composed of from view. Partially fill the interior with dry moss, and to this you can fasten everlasting and artificial flowers, mingled with grasses, grains and ferns, by first dipping the stems in glue, and thus produce the appearance of a basket filled with flowers, which, if artificially arranged, will produce a beautiful effect in the decoration. There are many more beautiful designs that could be suggested but it would occupy too much space. Many designs will however suggest themselves to those who are really interested in the work.

Abutilons.

Pendant flowers always possess a peculiar charm about them that other flowers do not, and the merits of the Abutilon make it superior to any. At least, that is my opinion, which is based upon my experience with this class of plants. It has all the good qualities required for ordinary room culture. The plants are of thrifty habit and grow in good shape with about as little training as any plant. The foliage, in form, bears a resemblance to the maple tree, and it is therefore known by the more common name of flowering maple. The leaves of some varieties are beautifully variegated, giving to them the appearance of mosaic work, while others are perhaps less attractive, yet all are clean, pretty and graceful. The Abutilon is almost a constant bloomer, and the pendulous bell shaped flowers are borne on long stems well out from the foliage, thus producing a delightful appearance as the the airy bells swing to and fro against the green and mottled leaves. The attractiveness is much due to their erect, stately form, and handsome foliage, and when kept clean by frequent syringing, a plant out of bloom is very pretty. The Abutilon is of easy culture, and will thrive in almost any soil, but it will do enough better if given a soil that suits it to make it worth 'while obtaining such. I find that it delights in a soil composed of loam and the turfy matter found beneath the sods of old pastures. To this add a very small quantity of sand and enough well-rotted manure to make the whole moderately rich. Provide good drainage and water regularly, and in sufficient quantities to moisten the soil all through. The Abutilon is quite free from the attack of insect enemies, although the green louse and red spider may sometimes be found on it. To rid the former, sprinkle or syringe with a decoction of tobacco stems, and the latter, by the free use of clear water all over the leaves.

If you want them to give you the most flowers during the Winter, it is best to set the plants in a cool sheltered place in the Spring, and keep them rather dry until Fall. About September cut them back well and repot them in fresh soil. Give water in small quantities at first, and increase by degrees as the new growth appears and the plant requires it. Aim to keep up a steady growth, and occasionally during the Winter it will do to give an application of liquid manure.

If you want to secure a bushy plant, pinch off the ends of the branches when they have grown to the height desired. This will cause new branches to start from every leaf joint, which can, in turn, be pinched back when they have made a few inches growth. By persevering in this process of pinching you can get all the branches you want, and, besides having a well formed plant, it will produce a large flowering surface. Some may prefer to grow them in the shape of a tree, and in this case but one stalk should be allowed to grow, thus forming a trunk, and when this has attained the height of three feet, pinch off the top to form branches. Allow no branches to grow, however, except those near the top, and by pinching these back as soon as they have made a few inches growth, you can have as compact and bushy a tree as you wish.

Of late years so much improvement has been obtained by seedlings variously crossed and by hybridizing, that the original species are nearly superceded. The varieties range through shades of orange, scarlet and rose to pure white. A variety known as Thompsonii has flowers of orange and scarlet while its leaves are beautifully blotched with bright yellow and light and dark green. The blotches of yellow and green are entirely distinct up to the point where each are joined, giving it the appearance of mosaic work. Those who wish to order some special variety will find the following variety of culture: Boule de Neige, pure white; Golden Fleece, pale yellow, and Rosaeflorum, rose color, one of the best.

Why not Train the Geranium?

It is seldom that we see a well formed geranium, while every where can be seen scrawny, ill-shaped specimens of all descriptions. A training of this plant for its own special beauty seems to be an undertaking that our flower lovers do not wake up to, but if all realized the signification of a well-trained geranium, I feel sure that there would be a great many more better formed plants.

A properly pruned and well grown specimen should secure to the cultivator, first a compact and bushy plant, instead of the long-legged, scrawny things usually seen. Secondly, a plant with leaves standing out on all sides, and not from one side alone, while the other shows nothing but naked branches. Thirdly, last, and perhaps I may add most profitably, a larger production of flowers, because they are produced from the ends of the branches, and of course the more branches, the larger surface there will be from which flowers may be expected.

To train a plant of this character, it must be brought under control at an early age, and carefully guided from its infancy. Take a good, strong, healthy slip with three or four leaves attached to it, and insert it in moist sand to root, first removing the head or top to form a cutting. When roots have started, remove to a two-inch pot, and buds will soon form young branches at the junction of each leaf, which after five inches should be stopped by pinching off the roots. Other shoots will then start from the branches already grown, and when these have made a good growth, they can in turn be pinched back. This will doubtless end the first season's growth, and the plant should be allowed to rest, simply giving it enough water to keep the soil from getting dust dry. The next season before starting it into growth, prune with patience, and they will be brought into exquisite form.

This should be done by degrees, depressing each a little every four or five days, and fastened with strips of long cloth, until they are brought into the right shape. Care should be taken not to bend the branches suddenly as they may break.

Repeat the operations already set forth, and especially the pinching process, and shifting to larger pots as the roots fill them, and with the proper management, if the plant is of a free growing variety, it can be grown to the size of four or five feet in diameter, with an annual production of several hundred trusses of flowers.

How different in contrast is such a plant with one allowed to grow to one or two stems, having a crooked side shoot. This is not an ideal sketch, for it has been produced in reality, and with better results than that portrayed above. Let our flower lovers try and be convinced. It will cost nothing.

Seasonable Hints.

Those who adorn and beautify their windows with plants during the long dreary months of Winter are surely decorators of taste and leaders in refinement. The culture of such is very simple, but we must not expect to grow plants to perfection unless some care is exercised in looking after their wants. The dust should not be allowed to accumulate on their foliage, because it stops up their breathing apparatus and gives the leaves a parched looking appearance. Spray or sprinkle the foliage frequently. This should be done at least once a day, and twice would be better. While the plants are confined to the ordinary living room, once in two weeks; or, better still, every week, sponge the leaves; or, in other words, wash their faces. This refreshes and raises a soft atmosphere, as does a Summer shower to vegetation. Water should be applied to the roots judiciously at this time of the year, because the sun has not so much power to evaporate the moisture from the soil as it had during the hot Summer months. The surface of the soil should get dry, but the plants ought never become so dry for want of water as to flag before applying it, and then enough should be given to thoroughly soak the ball of earth; discriminate as to which require more than others.

Few plants like a dry atmosphere, and a vessel containing

water, or even wet sponges, placed in the window near the plants, will continually furnish some moisture to the air, which is very beneficial. Do not water or sponge plants while the sun is shining on them, but let it be done in the evening or early morning; the evening I think preferable because the water has from twelve to fourteen hours to soak thoroughly through the soil before the sun commences its evaporation. Give an occasional application of liquid manure to Winter blooming plants, it will increase both quality and quantity of flowers, and, above all things, secure perfect drainage. More plants suffer for lack of good drainage than anything else.

On pleasant days admit plenty of fresh air, but not so as to cause a draft to blow directly on the plants. Keep the room in which you have plants as near an even temperature as possible; from 45 to 60 degrees at night is the safest general temperature—at the former there will be a handsomer plant growth but not so fine as if cooler, and by the latter more flowers. Keep down the aphis and red spider, the former by frequent application of tobacco water, and the latter by clear water.

Hanging Baskets.

Few things are moae graceful and attractive than a hanging basket with proper plants that have been kept healthy and in a growing condition; but a basket in which plants have become diseased and starved is about as sorry a sight as one will be apt to meet with; therefore, before entering into this subject very far, I wish to say to those who are not willing to go to the trouble of providing sufficient water and the attention necessary to secure perfect success, you better not attempt to keep plants in a hanging basket. The principal trouble and cause of failure with basket plants is due to the fact that they are apt to be neglected and suffer for want of water. It must be borne in mind that a basket is much more exposed to the upper and more drying atmosphere, and consequently evaporation takes place more rapidly than with an ordinary flower pot, and so requires a supply of water much oftener; but because of its not being as easily got at as plants in pots on the stand or bracket, it is too often given a mere "dash" of water without being removed. This kind of pretentious watering will not do, and in a short time ruins the effect of the plants, for the leaves begin to turn yellow and fall off.

To grow plants well in baskets requires a daily watering, and that thoroughly; a little given now and then without taking the basket down will not "fill the bill," and the plants must suffer. I find the best and most thorough way is to set the basket in a dish or pail containing water and leave it there until the soil is thoroughly wet, then it can be allowed to drain in the sink or any other receptacle, before hanging it up again.

A good wire basket is, to my mind, preferable to any that I have ever used, or seen my friends use, and this I line with nice layers of green moss from the woods, filling in with good rich soil that keeps it in place. It is an easy matter to take such a basket down and set it in a dish of water once or twice a week as occasion requires. This will give ample supply of water to the roots, as both moss and soil soak up all they can hold; and the moss will to some degree give off a continual moisture to the benefit of the plants.

As the number of plants in the basket is usually large for the quantity of soil it should be rich. What is wanted is a rapid luxuriant growth, and a good soil for the purpose may be made of about one part sand, two parts well-rotted manure and two parts turfy matter. There are quite a number of plants that can be selected for a basket, among which the Oxalis is perhaps the best. Its peculiar nature and graceful habit renders it especially adapted for this purpose. It varies some in habit, the leaves and flowers of the rose and yellow varieties droop more than the white, which holds itself erect, yet it is quite as beautiful and as well adapted for basket purposes, and their blooming qualities are about equally profuse in all Winter flowering varieties.

The best time to start this plant for window culture is in August or September, and if you begin with but one or two bulbs of a kind, they will soon multiply themselves many times.

Oxallis, like other plants, need a rest after they have done blooming, and this period will be indicated, toward the Spring, by their leaves turning yellow; they should then be dried off gradually in the basket and set in a cool dry place until Fall, when they may be reset and started into growth again, and the owner will soon be rewarded by an abundance of flowers for they are very free bloomers. The Ivy Leaved geranium is another fine plant and the richness and elegance of its foliage, and the drooping or trailing habit are qualifications it possesses, rendering it eminently serviceable for basket use. There are now so many varieties that one can indulge his or her taste in selection. Some varieties have leaves of green margined with white, others bronze or yellow, and still others are dark green, with a yet darker line about midway between the centre and margin. The color of the flowers range through rose, pink, scarlet and crimson to white.

Nolana is the name of another excellent plant, that is peculiarly adapted for basket use. It has succulent stems like those of the portulaca, and its leaves though somewhat succulent, are broad and fleshy and of a light green color. Like the little salamander, (as the portulaca is sometimes called) it will bear heat and drouth, and delights in a light soil and similar treatment.

The flowers are shaped sometimes like those of our Morning Glory (convolvulus major) only more firm. This plant will fully meet the wants of those who have but little time to devote to plant culture.

Tradescantia, commonly known as Wandering Jew, may be used to good effect, while Kenilworth Ivy, (Linaria cymbalaria)

Alyssum, and Petunia work in very nicely. A basket that I had last Winter, and which gave perfect satisfaction, contained for the centre a flowering Begonia, and outside of this a row of pink and white Oxalis, with Kenilworth Ivy for a border that drooped gracefully over the edge of the basket, and hung down in airy festoons; about each wire that suspended the basket a young ivy of the common variety was fast twining itself. While here and there a branch of Petunia reached out in such a manner as to give the whole a look of naturalness, and when all were in blossom they produced a beautiful effect, the colors blending harmoniously with each other, and with the several shades of green of the foliage. It was a miniature flower show, and won the admiration of all who saw it and it was a source of comfort and endless pleasure to the owner. Any one can have such a basket if they will but give it the necessary attention, which means sufficient light, water and air.

Rootings and Cuttings.

Most cuttings are readily rooted in moist sand. Where but a small quantity are to be promulgated the most convenient way, and one which I have used with perfect success, is to take an eight inch pot, plug up the hole in the bottom with a cork, and then fill about half full with clean sharp sand, such as builders use. Place a three inch pot in the centre without corking and press it down into the sand, until it is about even with the top of the other. Fill the space between the inner and outer pots with sand within an inch or so of the top, and it is ready to receive your cuttings, which should, of course, be inserted in the sand. Keep the smaller pot nearly filled with water, and place in a warm sunny position. If the bell glass is put over it the cuttings will root much sooner. As soon as the roots have formed, which will be indicated by new leaves or shoots, the young plants should be removed with care and potted.

Drainage.

Much has been said and written on this subject, and yet there are many people who do not see the necessity of drainage, and cannot understand why careful gardeners put all those crocks in the bottom of a pot. They imagine that by simply placing a plant in the pot, with an abundance of water and a high temperature, all the requirements of Nature have been complied with ; but, alas, the very means which are taken to secure their object lead only to disappointment. In growing a plant in a pot, we should remember that it is very differently situated than one planted in the garden. A potted plant can only have what is put in the pot, or what may be administered to it, and it has to accept the whole of what we put there. This is not so with the plant in the garden, for its roots can wander about in the soil, picking and choosing what it will take up, and what it will reject. Therefore, if a potted plant is provided with good drainage, it will have a source for discarding at least a portion of whatever is administered that the plant may not like.

Drainage is of the highest importance, and absolutely necessary for all plants. Without perfect drainage the overplus water cannot run off through the hole in the bottom of the pot as it ought to and if it remains about the roots it will soon cause the soil to become soured, the roots decay and the plants perish. The results of experiments have taught the careful man that plants will do no good if the soil in which they grow is kept too wet, and it is for this reason that prudent farmers and gardeners go to the expense of putting in drains through the land.

I would advise the principle of "an ounce of prevention is worth a pound of cure," and therefore take particular pains to see that potted plants have perfect drainage, because if a plant becomes sickly or diseased from imperfect drainage you will find it a difficult task to restore it to a healthy growing condition again, and often the plant will die from loss of vitality before the benefit of a change can reach its vital parts. If you err on the question of drainage, let it be too much rather than too little. My plan of providing drainage has given me perfect satisfaction, and I adhere to it strictly. I commence by placing a layer of small pebbles or broken crockery in the bottom of the pot, and over this put a little moss to prevent the soil from being washed down between the pebbles and stopping up the cracks. If at hand, I put a little fine gravel just beneath the layer of moss.

The Chinese Hibiscus

Is one of the most beautiful, showy plants among my collection. It is a woody plant, and quite a free bloomer. However, did it possess no merit as a bloomer, the bright glossy green foliage would entitle it to a place in every collection of ornamental plants. The flowers, which are very large, and of a rich red or crimson color, are produced on new growth during a greater portion of the Summer, and if the plant is grown in a warm, sunny and light situation during the Winter, it will bloom quite freely. It is of the easiest culture, requiring about equal parts of leafy mold and loam, to which add a little sand and manure.

The Hibiscus is also a desirable plant for bedding, as it succeeds admirably well bedded out during the hot dry Summer months. It is preferable, however, to use two or three old plants for this purpose, which should be well pruned and kept as dwarf as possible. In the Fall, when it is time to remove plants to the house or conservatory, care should be taken in lifting so as not to injure many of its roots.

For very large specimen plants, subject for decoration on the lawn, I think the best plan is to take an ordinary nail keg, saw it in two, and make the lower half into a sort of small tub, by fastening a hoop about it near the top, then bore the sides and bottom full of large auger holes. Set the plant in this tub, as you would place a small one in a pot, and plunge in the lawn to the rim. When the tub becomes filled with roots, the little rootlets will find their way through the auger holes in search of more room, and when it becomes necessary to remove the plant to Winter quarters, it will not be as difficult to lift as if planted out in the open ground, because the main roots will not be disturbed. Of course the roots that have formed outside through the auger holes may be cut off, but the shock to the plant will be very little as compared with breaking a main root.

The plant should not be allowed to become pot bound, and if its gets too large for the pot in which it is growing, turn it out and carefully reduce the ball one-third, cut the branches back well, and repot in the same size pot, with fresh soil, and water carefully until well established again. It is best to repot in the Spring, about the middle of May.

The Hibiscus is readily propagated from cuttings inserted in sand, and, by repotting the young plants as often as required, fine specimens can be produced in a few years.

The aphis or green fly sometimes attacks the Hibiscus, but it is easily rid of them by the use of tobacco, either in smoke or infusion.

There are quite a number of varieties in the trade, some more or less double, and some with various shades of crimson, red, and yellow, to almost white, but I doubt if any of them are entitled to rank more than as mere descendants of the Chinese Hibiscus. Those who would like to order some special variety will find the following worthy of culture: H. miniatus semi-plenus, vermillion scarlet. This variety is a sort of semi-double, producing petal-like bodies on the stamenal tube, and is very showy. H. dennisonii, white, shaded with very light rose or pink. H. fulgidus, flowers very large, carmine scarlet, changing to deep crimson at the base of petals. H. sub-violaceous, double, clear carmine, tinted violet. H. cooperii, tricolor, has beautiful variegated foliage and crimson flowers.

Dicentra Spectabilis.

Its common name is Bleeding Heart, and it is a handsome plant, bearing flowers similar to those of the Alleghany vine, only they are far prettier. The white and bright rosy pink flowers hang gracefully in racemes from the bending branches, like so many heart-shaped gems, and present a truly beautiful sight as they are seen among the sprays of foliage. This species of Dicentra (spectabilis) is a hardy perennial, and is supposed to have originated in the north of China. It makes a very fine plant in a few years, and its own merits and attractiveness should place it among every collection of perennials. Its requirements are very simple. Keep the soil mellow and clean about the plant, and during Spring work in a little fine manure. In the North, where the

Winters are severe, it is well to protect the roots by a light covering of leaves and litter. The roots may be allowed to remain in the same place for several years, where they will spread in all directions and form a large clump, that will produce a floral display unequaled by few of the floral world.

Any of our readers who have seen this plant from childhood may think the above somewhat extravagant praise, but the real lover of the beautiful who may perchance set eyes upon it for the first time will find it difficult to produce their admiration in words.

I have been informed that Dicentra spectabilis will do well in the house or conservatory, and become a fine specimen, but I have never tried it, and of course cannot recommend it as a house plant. However, should any of our readers have occasion to try the experiment, or if any have grown it in the house, I would be pleased to receive a full account of it, and will publish it for the benefit of others.

Browallia.

It is an annual and a native of South America, and well grown plants will bloom quite freely. If a succession of flowers are desired, it is well to sow seeds at intervals during the Summer and Fall, so as to bring the plants in bloom at different times of two or three weeks apart. The seed is very fine and takes its own time in germinating, so do not be discouraged if it is slow in coming up, but allow plenty of time, and when the seedlings make their appearance, shade them from the very hot sun until well started.

For house culture, it is best to grow them in pots, and if three plants are set triangularly they will form a fine mass, and be more effective than if grown singly. Strong stocky plants can be produced by pinching back the branches as they increase in size, thus causing them to put forth side branches, which they will do by this means very generously.

Among the varieties of Browallia we have, B. elata grandiflora, entirely blue. B. Cerviakowski, blue with a white centre, and there is also a white variety called B. elata alba.

Ageratum.

Another fine plant for the Winter months, that bears light blue and white flowers, is the Ageratum. The seed should be sown at the same time as the Browallia, and the plants require about the same treatment. A light, rich soil, composed of equal parts of leaf-mould and loam, with a little rotted manure, is a suitable compost for them. There is a new variety of Ageratum recently introduced, called Swanley Blue, that is dwarf in habit and has flowers of a deep blue. Now that we have entered into the subject for the Winter, and have given our correspondents a fair list of blue sorts from which to choose, I will go a little farther and introduce some white flowers of about the same general class, for the benefit of our readers who may possibly prefer white in the place of blue, or perhaps some would like both. I will first say, however, that there are white varieties of both Browallia and Ageratum, that are equally as good as the blue, and thrive quite as well, but I have two other plants in mind, about which I wish to say a word—the Stevia and Eupatorium. These plants are similar in their general structure, belong to a large order of the floral kingdom, known to botanists as Compositæ, and bear a close relationship to the Ageratum. The flowers of the Stevia and Eupatorium are valuable for bouquets; they are quite small, but the umbels are so large and full that they produce a very showy appearance, and a well-grown clump of the variety, Stevia compacta, will produce a beautiful sight at the holiday season.

Of the many species of Eupatorium that grow in this country, E. riporium, and E. elegantissimum, are the most suitable for late Fall and Winter culture. Well-grown plants, with proper care and a temperature between 55 and 65 degrees, should bloom profusely during a greater portion of the Winter months.

Floral Brevities.

A little more care than usual should now be given in protecting plants from cold snaps, especially in rooms that are not supplied with constant heat. On very cold nights it is well to remove the plants away from the window several feet and cover them with some light article. Newspapers are good for this purpose, but the "patent plant bed cloth" that is made by U. S. Waterproof Fibre Co., N. Y. City, is better, and it will pay to keep a few yards on hand, for it is a very convenient article to have.

It is just the thing to protect plants in the bed from light frost, that often visit us northerners in the Fall, when we are unprepared.

Cut flowers, if not badly wilted, may be made fresh by cutting off the ends of the stalks and putting them in real warm water, (stalks and leaves, but not the flowers) for an hour.

I have often wondered if my floral friends, while working among their plants, ever stop to think of the invalid who is unable to enjoy the pleasures derived from cultivating a few flowers. We ought not to forget to give a share of all our flowers to the sick and sad ones, for it is by such ministrations that we may help and encourage them.

Spring Preparations.

About this time, as the almanac says, we get the first intimation of Spring. The lover of flowers will have plenty of work to perform, such as making plans for the arrangement of beds as soon as the weather will permit. The advantage of having strong young plants in readiness at that time is so great that various means are resorted to in order to accomplish this end.

While it is true there are some seeds, such as Balsams, Pansies, Marigolds, and Zennias, which may be successfully germinated by the amateur, yet the majority are somewhat difficult, and usually require experience and care. However, I do not think the cause of a large number of failures with seeds is due so much to want of care, as from lack of knowledge as to the best and most successful methods of sowing them, and their delicate constitutions.

At the season of the year when the seeds are usually sown, there is less likely to be a succession of sudden changes in many parts of the country—warm showers, followed by a cold spell, and perhaps a slight frost, enough to kill many of the tender kinds, while cold, drying winds will ruin many more.

Besides all this, the shortness of our seasons at the North makes it desirable to start seeds early in order to have early blooming plants. Now, to overcome these natural difficulties, the florist, after repeated experiments, has found that a good propagating house, supplied with boiler and pipes heated with hot water, is the most reliable way of starting seeds. However, this is too expensive for the average flower grower, with a small garden, and where a hot bed cannot be used, the amateur has contrived a plan of propagating them in boxes in the house. This is the simplest plan, and one within the reach of us all, so we will give a few hints upon the subject that may prove useful.

After obtaining a shallow box, the next important requisite is to secure a good porous soil that will not bake or get hard. For this purpose I have found nothing better than equal proportions of sand, loam and leaf mould, and with this fill the box nearly to the top; then sift enough through a rather fine sieve to fill the box nearly to the top, level it with a straight stick, and jar the box sufficiently to settle it. This being done we are ready to plant the seed, which should be sown in rows, and as a general rule about the depth of twice their thickness. Label each row properly, cover with sifted earth, and make it level and firm with a small, smooth board.

In applying water care should be taken not to overdo it at the beginning, then cover with a moist cloth, or, better still, sphagum moss, such as florists use for packing flowers in, and keep the box in a moderately warm place until the young plants break through the ground, then remove the covering and put the box in a cooler situation and give them plenty of light, or else the young seedlings will come up spindling and "leggy." On mild days, when the weather will permit, admit all the air possible, and sometimes they may be put outdoors for an hour or two during the middle of the day.

At the appearance of the third and fourth leaves it is time to transplant them, and if the weather is such that they cannot be removed to their permanent beds for several weeks, plant them an inch and a-half or two inches apart, in boxes of good rich soil, so that they have sufficient room to make a good healthy growth before their final removal to the garden. Amateurs, if you really care for flowers enough to study their habits and needs, with a little attention to the suggestions here given, you may succeed in starting plants from seeds.

Sweet Flowering Peas.

Among our select Summer flowers there are few more desirable than the Sweet Flowering Pea. Every curve in its peculiar form is graceful, and presents a charming individuality that wins and holds the admiration of all true lovers of the beautiful. They give us all colors, from the darkest pink to purest white, and include the darkest purple imaginable. The beautiful blossoms send out their delicate fragrance, perfuming the air, and invite attention by their rare loveliness.

Looking at the sweet pea from a botanical standpoint, we find that the petals are five in number, and, from their peculiar arrangement, have received different names, and are grouped into pairs with the odd one standing nearly erect back of them. This uppermost petal is called the banner, the two lower petals, lying in close contact, are called the keel, and the two horizontal petals enclosing them are the wings. All this will be clearly understood upon seeing one of the flowers. The pea is a representative of a very large order of plants known as the pulse family, the scientific name of which is Legaminosæ. This long word is derived from the Latin lego, to collect, meaning that the seeds are collected in a pod called a legumne.

The sweet pea delights in a cool, moist (but not wet) situation, and good rich, mellow soil, which should be well worked as soon as the weather will permit. There should be no waiting for warm weather or the first of May, for they need considerable time to sprout. Use a generous supply of the best seed that can be produced from some reliable florist, and sow them about four inches deep. In planting the seed, a very good way is to sow two rows about twelve inches apart, so when staking time comes they may be trained to one row of stakes, set between the two rows. Another method is to mark out a circle two feet in diameter and sow the seed around the edge of the circle, then put a barrel hoop just inside the circumference and fasten it down to the ground with pegs. Now drive a stake in the centre, leaving about four feet of it exposed to view, and from the top of this run strings to the hoop for the vines to climb on. When the peas have come through the ground, keep the soil well stirred and pulverized until the season is well advanced, then mulch with some light litter to keep the soil cool, and if the dry weather (which we usually get at this season) is too severe, keep the ground moist by irrigation.

Sweet peas, the embodiment of all that is beautiful and fragrant, variously colored and tinted like the Pansy, seems gratefully to acknowledge our admiration in a silent but understanding way, by doing best when their flowers are freely cut.

A Renovated Garden.

In the Spring time, after the snow has succumbed to the sun's powerful rays and mother earth is again unveiled, it is then that we behold many a yard littered with a confused museum, so to speak, of smashed boxes, old fruit cans, crippled barrels and a variety of unsightly truck that is small enough to be deposited there. Of course, much of this degenerated nuisance will doubtless be removed as the Spring advances; but quite often there are other objects, which are really quite as unsightly, that are overlooked. A short stroll in the country, through the streets of small villages and towns, and large cities, will at once convince the reader of that fact, and the following illustration of a place I have in mind will, with some slight variation, be applicable to many others.

In front of the premises may be seen an old rickety fence, nearly ready to fall down, and several pickets gone, and others broken and battered or only half remaining. A corner of the yard may contain a shrub (usually a dear old Lilac bush) which at one time was no doubt a beautiful sight, and may have added some beauty to the place or, to use a more expressive phrase, it may have taken off much of the curse with which such places are too often abounded. However, this bush seems to have been given the same care as that bestowed upon the fence and the result is a poor, miserable, half-wild scrub with many decayed and broken branches. An old archway may often be seen leaning against the house as if for support, and over this archway and trellis a grapevine is usually allowed to ramble at will, for as a general thing, grapevines grown in such places are seldom cultivated, pruned or laid on the ground during the Winter and protected with a light covering. If there is any space at the side of the house or in front, it is usually filled with a large display or overrun with tall weeds and rank grass, each apparently bent on crowding the other out and taking full possession. Now, anyone with a little taste for gardening can turn such a delapidated place into a beautiful little paradise with but a small outlay of time and money. I have often noticed that a well-kept small place with a general use of flowers has a good influence on the community, especially in its immediate vicinity, and is often a beginning of its general improvement, for apparently some persons only need an example of what a little care and attention will do toward making home attractive to adopt similar improvements in their own places, and when this is once begun it is quite likely to be continued, for the love of flowers will grow, and enjoyment derived from them by the true lover is so full of fascination that it seems never to cease.

It is not proposed to set forth a plan for renovating neglected places from which no departure can be made because what may suit one person may not at all be satisfactory to another, but it is my design to offer hints and suggestions on the improvement of the place here outlined by way of assistance to those who under similar circumstances would like to improve their homes. In place of the old rickety fence, a new one could be constructed, or the old one repaired, or in fact, it may be done away with entirely. However, if a front fence is preferred, I would advise training a vine of the Ampelopsis Vetchii (American Ivy) over it, as it would give perfect satisfaction. It makes a luxuriant growth, and during the Summer is completely covered with a dense mass of beautiful, bright glossy green foliage, that turns to crimson and scarlet of every hue and shade during Autumn, at which time it is grandly beautiful. Our people did not at first appreciate this vine, and until the past few years not many of them were planted, yet no one could pass through the woods and fields, especially in the Fall months, without admiring its great beauty.

The old bush in the corner, if like some I have seen, should be removed and another set in its place. The kind of shrub or tree used will of course depend upon what is most admired or would produce the best effect. Hydrangea, Spirea, Syringa, Snowball, Wagelia, Lilac and Honeysuckle, are all desirable shrubs, while some of our evergreens, Silver, Birch, Maple and Tulip Tree (Lirodendron) are among the desirable trees. The latter is a beautiful native that is seldom noticed by our tree planters, but why I know not. It is seen in European ground more frequently.

As regards the old archway, I think it would improve the appearance of the old place if it were removed entirely, and a vine or climbing rose set close enough to the house to allow of its being trained thereto. For this purpose I know of nothing better than a choice from this list: Clematis jackmanii, Wisteria (apois tuberoes), hardy climbing roses; Baltimore Belle, nearly white, pale blush; Queen of the Prairies, rosy red; Boursault, dark crimson, and sometimes Honeysuckles may be used to good advantage. Two of the finest varieties are Hall's Halleana, leaves clean and handsome, flowers white and fragrant, and Belgian, or monthly fragrant, red and pale yellow, very fragrant and free flowering. If there is a porch in front or at the side of the house it may be filled with climbing roses, or where a dense shade is desired, use Ampelopsis Veitchii.

It is nice to have a garden well filled with vegetables, so that they may be gathered fresh for home use; but it is unnecessary to expose this part of the garden to street view, as seen in many places, when a row of grape-vines, planted across the yard about fifteen or twenty feet from the fence and parallel to it will, in a year or two, make a fine arbor; and, with a little attention, bear an abundance of fruit, and at the same time obstruct a street view of the back yard. Just back of the grape arbor would be a good place for raspberries, gooseberries, currant bushes, or a small bed of strawberries, depending upon the amount of ground at hand and taste of the owner. In place of the tall weeds and rank grass a good lawn should be prepared, and the grass kept short by frequent cutting. It will add materially to the appearance of the place. Neatness is one of the charms of the yard, and the cultivation of the garden, the ornamental planting of grounds, and the free uses of flowers are marks of progress, and a proof of refinement as well as the highest state of cultivation.

If my young readers were encouraged and could be persuaded to take an interest in the renovation of the garden and cultivation of flowers, I feel sure that it would be a step toward producing an excellent moral effect upon their young ideas, while at the same time it would be a help to their parents.

In conclusion, I will offer a few suggestions in regard to the flower garden and then leave the reader to decide which of the places, here described or outlined, is worthy of the name home.

About the various styles and methods of arranging beds and flowers, I will say nothing, because tastes differ, and it is best to let each one suit him or herself in this respect. However, I would advise planting each kind by itself. You will find it the most satisfactory plan. A bed containing a half dozen kinds mixed is never so attractive as one in which there is but one kind.

About the centre of the grass plot in front of the arbor is a good place for a geranium or foliage bed, while clumps of roses, pæonies, lilies, etc., may be grown here and there on the lawn with good effect.

I earnestly hope that any lover of the beautiful, who may have neglected his or her garden, will find some hints or suggestions in the preceding lines which will be of assistance, and that you will set to work and take hold of your gardens in earnest and show your neighbors what you can do. If you should meet with failures do not give up, but try again.

Seasonable Hints.

All work this month should be pushed forward, and vacant beds should be in order to receive plants intended for them, and if the ground is in good condition and the weather is mild and settled, about the 18th or 20th, no time should be lost in bedding your plants out. Take particular care now to keep your window plants well supplied with water, because the sun is becoming powerful and evaporates the moisture more rapidly. Calla lilies that have completed their flowering season should be allowed to rest, and may be turned on their side in a cool, shady place until Fall; but they ought to be looked at occasionally to see that the soil does not become dust dry. Seeds of annuals may be sown in the open ground about the 20th, if the weather is settled. Now is the time to take cuttings for next Winter's flowering.

Tree Planting.

This is a subject that grows in interest, and demands more attention every year, for the reckless manner in which trees have been slaughtered during the past few years, would lead to assume that they were man's enemies. I will admit that they were an obstruction to the progress of our early settlers of this country; but their reckless destruction of late years is carrying the matter to the other extreme, and our people are just beginning to realize the fact. It has been scientifically demonstrated that the changes of climate, increase in violent storms, land slides and inundations, causing destruction and death all about us, are due, more or less, to deforesting hill and mountain sides, and unless this progress is arrested there will soon be nothing but "bald knobs," down which the falling rain may rush in torrents, carrying with it all the fertilizing elements from the soil into rivers, lakes and oceans.

In some parts of Switzerland there is a law forbidding the destruction of a tree, without planting another to take its place. This is the outgrowth of necessity, for the trees have gradually disappeared and not many now remain, excepting on the high slopes of mountains, and those are of inferior size and quality. That trees are now being planted I will admit, but yet—to every newly planted tree—thousands are being cut down every year, to say nothing of the acres that are scorched and blackened by fires, and a fact beyond my comprehension is that for miles along some of our railroads, one can see nothing but trees and shrubbery on either side, while many hillsides at a distance present a nude appearance. If land is desired for cultivation or wood for fuel let it be along the railroads; but let the mountain and hillsides, rocky places and rivers, remain covered with trees where feathered tribes may make their homes and chirp sweet notes of praise, and where beneath their shade nature's beauties bloom in the most cheerful effects.

It is unquestionable that more tree planting is desirable in nearly all parts of the country; it certainly is in the Middle States, and Arbor Day is an outgrowth of this popular conviction. The design of an Arbor Day is a good one, and the rapidity with which its observance from a feeble beginning has developed, is the most direct and formal expression of the public toward the subject of forestry. To avoid the possibility of misunderstanding, I would say, however, that it is not designed to be a day for the actual work of setting out trees. It would not be advisable for the people in all parts of a State to unite on the same day for that purpose, because the season of transplanting trees in the southern part would very likely be too early in the northern part, and vice versa. Now to overcome this as well as other objections it is intended to hold Arbor Day as a day of celebration, when the people in each community, both young and old, may assemble and discuss the importance of American forestry, the manner of acquainting the young with the laws relating to tree growth, their care and value. Reports of the tree planting during the past season may be read, and individual notes and methods for comparison presented to the members of assemblage, as well as general plans for future work. These meetings should be enlivened by appropriate exercises, accompanied by short speeches, songs and music. In other words, it should be such a day of celebration and enjoyment that all would gladly welcome its return.

But Arbor Day means still more. It is a reminder that home with its enjoyment, beauty and comfort can hardly be considered as such without that air of refinement and good taste which trees, shrubs and vines impart to it. Nor is this all, for the impression that can be made upon the young people will doubtless cause the rising generation to become more enthusiastic in tree culture than those preceding it.

The introduction of Arbor Day into many of our public schools is worthy of the efforts made to engage the pupils in its observance, for the practical outcome may already be seen in the improved appearance of the school grounds, from plain, cheerless, uninviting school yards to attractive shady parks and groves. In the schools of one of our States where Arbor Day exercises were observed last year, the students enrolled with hearty co-operation and the exercises were opened with music and song. The president of the meeting, who had previously been chosen, then read the Governor's Arbor Day proclamation and delivered a short speech, after which brief and appropriate addresses were then made by some of the teachers and principals while the students cheerfully varied the program with songs, poems, music, recitations and declamations. Several of the leading men of the community, by invitation, also made short and pleasant speeches, expressing sympathy with the Arbor Day movement, each speaker treating the subject from his standpoint. The day was expressed as a most enjoyable one, and its return will doubtless be welcomed with delight and increased enthusiasm. From this brief explanation it will readily be seen that what is now needed is the entire approval and support of every school and college in this country. Let every one be interested in this subject. It may perhaps awaken a new thought in some not already interested.

Unique and Curious Plants.

The Rose of Jericho is a very curious plant that grows in sandy places in the vicinity of the Mediterranean Sea, being a native of Arabia and Egypt. In its dry state the branches incurve, assuming an oval or almost round appearance, and in this shape it is carried for long distances by the wind. By some superstitious people it is believed that the plant expands its flowers every year on the very day and hour that our Saviour was born. The Rose of Jericho is known botanically as Anassatica Heirochuntica, and is so called because after drying up or assuming a dead appearance by placing it in water for a short time the plant will resume its original shape.

The Air Plant, with its long botanical name (Tillandsia Utriculata) is a native of Florida, and belongs to the pineapple family, which plant it closely resembles when not in fruit; but unlike the pineapple it requires no soil, and in its native haunt may be found growing in the tops of lofty trees, where it derives all nourishment from air and water. The peculiar manner of growth of the air plant enables it to retain nearly a pint of water at a time, while at other times it may remain dry for a long period without any apparent injury. A full grown plant will measure about twelve inches in diameter, and upon attaining full size will bloom, and the effort invariably kills the plant, and as soon as the seeds have matured it dries up and falls from the tree.

Fuchsias.

There probably is no more popular flower under cultivation than the fuchsia, or ladies' eardrop, as it is frequently called, unless possibly it be the geranium, and although it is not one of the oldest flowers of our gardens and windows, still on account of its graceful habit and varied, yet delicate coloring, it has become deservedly popular, and when, added to this, we consider the fact that it is one of the best plants of the floral kingdom for pots or boxes on shaded verandas, or for beds in partially-shaded places in the garden, there would seem no reason why this popularity should not continue. During the rage for something new a few years ago, it was somewhat overlooked, as were others of our old and meritorious plants. But sooner or later there will be a change, when our old and neglected flowers will become the popular plants of the garden again, and then the Fuchsia will rank among the best.

The Fuchsia is of American origin, having been first discovered in Chili, South America, where it grew in moist shady places. Species have since been discovered in different parts of the world; Mexico claiming one and Australia another. The generic name, Fuchsia, was given in honor of Leonard Fuchs, a noted German botanist who, it is understood, was the original discoverer. Some of the varieties, now under cultivation, would scarcely be recognized as members of the same family at all, by those who are only familiar with the ordinary garden sorts, and though these original types are more interesting from a botanical than a floricultural standpoint, still they are occasionally grown, and make quite beautiful specimens. F. Fulgens, a South American species, has a tuberous root like a dahlia, only much smaller of course; the leaves are quite similar to those of the more commonly grown garden varieties, while the long tube-like brilliant flowers droop in clusters from the ends of the branches, and altogether make quite an attractive plant. F. Coccinea is another tuberous variety, but seldom seen among general collections or advertised by florists; but F. Procubens, a native of New Zealand, has become quite popular as a basket plant, and being of a trailing or drooping habit, does very well for that purpose. The peculiar shaped flowers of a bright yellow, brown and green color, beautifully blended together, are followed by berries that become bright red when ripe. These comprise the peculiarities of some of the original types, which were brought under cultivation, propagated and crossed to produce hybrid varieties, and so changed that now we behold them arranged in a garb of the most charming beauty.

Some persons seem to be impressed with the idea that Fuchsias are Winter bloomers, and do not make much of an effort to produce flowers during the Summer, hoping in so doing to be rewarded by a few flowers during the Winter months. Those who follow such an idea, however, do not realize what beauty and pleasure the Fuchsia is capable of producing under the most favorable conditions, for it is properly a Summer bloomer and blossoms freely at that season with about as little care as any plant. When the requirements of this lovely plant are are once understood, the cultivation is quite simple. It delights in a rich porous soil, good drainage, judicious watering and plenty of light, but not strong sunshine.

Young plants, if started early and properly cared for, may be brought into flower the first season, but I do not consider them as satisfactory as well developed two or three year old plants having good strong roots and branches. If, however, it is desired to begin with young plants, cuttings may be readily propagated in moist sand ; an operation that will prove most interesting and instructive, as it affords a good opportunity for observing growth and development. A simple, yet inexpensive, propagating case may be formed by the use of a small box half filled with sharp sand. Plunge a small flower pot in the center to receive the water supply, cover the whole with a light of glass and place the box in a sunny position. As the moisture evaporates, fill up the pot with water and it will soak into the sand through the hole in the bottom without disturbing the cuttings. Roots will usually form in the course of two or three weeks according to conditions, and the young plants should then be carefully potted in a fairly rich soil, composed of leaf-mould. Nothing seems to suit them quite as well as earth from about the roots of trees and old stumps in the woods and pastures, where leaves have drifted and decayed, and for young plants leaf mould will be quite rich enough without the use of fertilizer, but for large plants of two or three years' growth, well rotted manure may be added in small quantities as the condition of soil requires it. Be sure to provide good drainage, or you will have no end of trouble and disappointment with a sour soggy soil. Water only when the soil appears to need it; then apply a sufficient quantity to moisten the earth all through.

The Fuchsia is a desirable plant for a window or veranda box in places not exposed to the strong sunshine of mid-day, but if it is preferred to keep them in pots, they may be plunged in boxes of soil during the heat of Summer as a means of keeping the roots cool and lessening the evaporation of moisture that so rapidly takes place when the pots are exposed to drying winds and a warm atmosphere.

It is important that the plants be shifted as often as the soil in the pots become filled with roots, in order that they may not get root-bound. Should the ball of earth be matted with roots, beat around it gently to loosen the crust to allow new roots to start out.

Nearly all plants are more or less troubled with insects of some sort, and the most destructive enemies of the Fuchsia are the red spider aphis. The red spider is very small, and upon close examination appears like grains of cayenne pepper. Whenever the leaves of your plant seem troubled, carefully inspect the under side of them, for that is where the spiders generally do the most harm. A hot, dry atmosphere is their delight, while moisture is sure death, and clear water forcibly applied to the leaves, especially the under side, will rid them. Aphides, or green lice, are larger and more readily observed, but do not infest the Fuchsia as frequently as many other plants, yet when the conditions suit, they may be found there. Sulpho tobacco soap will exterminate them if persistently used. It can be had of most florists and seed establishments.

The varieties in Fuchias have increased very rapidly during the past few years, and nearly all have more or less good qualities, but the list here given comprise only such as seem to possess the most merit. The varity Speciosa, is, all things considered, the most satisfactory of any. It is is almost constantly in bloom, and may be counted as the only true Winter bloomer of real merit. Of course, there are others that may be brought into flower during the Winter, if allowed to partially rest during the Summer, but they are not as floriferous as Speciosa. However, of the most desirable for this purpose, I would select Mrs. Marshall, having white sepals and carmine corolla; Carl Halt, crimson corolla striped with white; Pearl of England, sepals white, corolla rosy-scarlet; Arabella, tube and sepals waxy white, corolla rose; Rose of Castle is a fine old variety, having bluish-white tube and sepals, and a purple corolla, and Storm King, a very free bloomer, with a pure white corolla.

Among the most desirable sorts for Summer flowering, Black Prince takes the lead as the most floriferous variety. It makes a strong and upright growth, branching freely, and producing its flowers in great profusion. A well grown plant three feet high, in full bloom, is a minature flower show all by itself, and will last for several weeks. Tube and sepals bright carmine; sepals broad with green tips; large open rose-colored corolla.

Elm City is a grand old variety with a double purple corolla, and deserves a place in every collection. Madame Van der Strass is considered the finest double white yet produced. It is quite as floriferous as its rival, the Storm King, and is much superior to that variety in habit of growth. Mrs. E. G. Hill is a vigorous upright grower of considerable merit, with very large, double flowers of a crimson and white color. If you want a very large Fuchsia get the Phenomenal, which is unquestionably the largest variety grown, and notwithstanding its immense size, the flowers are beautiful and very double. Convent Garden will produce a graceful effect if allowed to droop naturally from a central support, instead of being fastened to a stiff awkward-looking trellis as generally seen.

Callas.

The Calla is a native of a warm country where it grows rampantly in low murky ground, and if the most satisfactory results in cultivating it are to be obtained, we should strive to make the conditions as near like those under which it thrives in its native habitation.

In our Northern States the Calla is almost universally cultivated as a Winter blooming plant, and should therefore remain practically dormant during the Summer months. After blooming, the plants may be kept growing until Spring, though not as vigorous as before, then gradually withhold water so that the bulbs can mature their growth, until only enough is given to prevent the soil from becoming dust dry. When the weather is warm enough to allow removing the plants out of doors without injury, place them in a shady nook, with the pots partially on one side, resting the upper edge on a notched block of wood. In this position they will receive only a portion of the rain that falls, and this will generally keep the soil moist enough without watering, unless there should come an unusually dry spell, in which case, a light sprinkling will suffice. About the 1st of July it is a good plan to turn the plants out of their pots and set them in the garden border where they will be somewhat shaded from the direct sunshine of midday, as the sun is usually very powerful and burning at this season. Here they may remain until it is time to repot them again in August; the idea being to keep the plants in as nearly a dormant condition during the Summer, as is consistent with health, and thereby afford an opportunity to store up strength for Winter forcing and blooming.

Toward the latter part of August or early in September is the time to repot callas, if wanted for mid-Winter flowering. Provide good drainage and fairly rich soil, composed of black muck from a swampy, if it can be obtained, working it in thoroughly. Carefully lift the bulbs from the garden, and remove all the soil before potting, after which thoroughly water, and place the pots in the shade until the bulbs are established. As the evenings begin to grow cold, it is a good plan to bring the plants to the veranda, or other convenient spot, for shelter, so long as there is no probability of frost. When it is necessary to house the plants, remove them to the coolest room, admit fresh air daily, as long as the weather will permit, and they will become accustomed to the change so gradually, that it will apparently have no effect upon them. Of course, the pots should set in a deep dish or pan, as the plants require a large amount of water, and especially so when growing. Let the water be quite warm, pouring it slowly on the surface soil so as not to injure the roots, and when flower buds are expected, (which will be indicated by a swelling of the main stalk), do not change the plant from one room to another for the sake of having them in a conspicuous place, as a difference in temperature is likely to destroy the bud or flower. If a bud seems to develop slowly, I find that a little F. F. F. F. aqua ammonia applied in the proportion of a tablespoonful to a pint of water is of much benefit; using the same about once a week. Sometimes calla blossoms are inclined to take upon themselves what may be termed "freaks of nature," by coming out in a garb of green. This green spethe is a partial reversion to the original leaf form, or bract. A bract is generally green and in the form of a leaf attached to the base of a blossom from which the flower arises, but in some cases they assume peculiar forms and colors like flowers. Such is the case with the Calla; its large white envelope or spethe, commonly spoken of as the flower, may in a descriptive sense, be considered as the bract changing to a flower-like form. The real flowers of the Calla are situated upon the oblong central column called the spadix, covering its surface entirely.

Pinks and Carnations.

The Pink is the familiar representative of a natural order of herbaceous plants, botanically termed caryophyllaceæ, but perhaps better known as the pink family. In its wild state, the Pink is found growing on the south side of the Swiss Alps at a low latitude, where the Winters are not severe, and although a great favorite in grandmother's garden, it was but little known prior to 1772, when a gardener to the Duchess of Lancaster, named James Major, was fortunate enough to have a seedling plant produce a double flower with laced petals. Mr. Major, grasping the opportunity, succeeded in propagating a stock of this new variety, offering it for sale to the public with marked success, and from that period the Pink made a rapid advancement, until now we find it large and full, with exquisite forms, handsome colors and varigations, and with lacings as perfect as can well be conceived. There are two classes into which we may properly divide this plant, but the varieties seem almost unlimited. The ordinary garden Pink, containing the perennial class, dianthus plumarius, and the biennial, or old and well-known Chinese sort, D. Chinensis, of which many new and superb varieties have been introduced during the past few years. D. Caryophyllus, the well known and much esteemed Clove Pink, is the parent of a double-flowered class called Carnations and Picotees, which are well adapted to pot culture in the window and conservatory. As a matter of fact, the Carnation is one of the sweetest flowers we possess, and its neatness, beauty and fragrance, together with the great variety of colors, markings and free blooming qualities, form an array of good points that have caused it to become a leading favorite with many of our flower lovers. There is a very nice arbitrarial distinction of colors, made by florists, between the Carnation and Picotee, which, to the average amateur may not be very clear, but a short description here of each, will, I hope, render the matter easily understood so that they may be clearly defined The petals of a good Carnation must be firm, smooth at the edges and have broad stripes of color running through from the center to the edges of petals. Florists' varieties are divided into six classes, known as rose flakes, with bars of rosy shades on white ground; scarlet flakes, with purple bars, selfs or cloves, in which one or more colors are distributed evenly all over the flower; bizarres, with scarlet, purple and white bars, and any three or more colors traversing the same way, varying from light to dark, and fancies, which include flakes and bizarres of unusual colors or neutral tints. The petals of a Picotee also have smooth edges like the Carnation, and virtually differ only in the arrangement of color or markings. Picotees have a solid ground of light or white color; sometimes yellow, and are usually ornamented with a narrow band evenly penciled more or less heavily around the edge of each petal. They are always chaste and beautiful, and being alike in habit and hardiness as Carnations, require similar treatment in raising them.

Pinks are considered perfectly hardy by florists in general, and so they are, but nevertheless I find that it pays to protect them lightly in latitudes where the Winters are severe. They are of compact growth, with narrow foliage resembling grass to some extent, and are easily grown in any good garden soil. They may be propagated by a division of the root, pulling off single tufts with short "heels" from the main stem and planting them closely together in a shady position. This should be done just after flowering, and when well rooted, the young plants may be transplanted in a permanent bed. If grown from seed for early blooming, it should be sown as early as April under glass, in good, mellow soil, and treated as other seedlings.

The Carnation may be propagated by layering, or from cuttings in moist sand, and should be started between March and May, if wanted for Winter flowering. As soon as the weather becomes settled and the ground can be worked, bed out the young plants for the Summer in a light place, where they can have plenty of air and sunshine; making the soil light and fairly rich, to a good depth. About mid-Summer they will be inclined to blossom, when the flowering stalks should be cut back, and the process repeated as often as the buds appear, up to the time of potting in the Fall; the idea being to produce a steady, healthy growth during the Summer, with numerous strong, compact and vigorous branches, when it is time to pot them. The potting should be done toward the end of August or early in September in the latitude of New York, always supplying a liberal course of drainage material. Shade the newly potted plants for a few days until they become established. Before potting, it is a good plan to run a knife blade in the ground around each plant at a reasonable distance from the center to cut the long roots that would otherwise be crumpled up in the pot. If the plants are allowed to remain thus for a few days before lifting, the cuts will heal and new rootlets form, leaving the effect of removing and transplanting barely noticeable. The Carnation is subjected to attacks of red spider and aphis, but by frequently syringing or sprinkling with clear water for the former, and an occasional application of sulpho-tobacco soap for the latter, they may be kept pretty free from these insect pests. The tobacco soap may be had at most seed establishments.

During the past few years some remarkably large flowering varieties, with most exquisite colors have been introduced, and although it would seem almost useless to even think of further improvement in the perfect strain of carnations now grown; yet we can never forecast the time of a new departure in nature, and some day there may be quite a revelation in the advancement of new types.

Parsies.

Who does not know the Pansy (whether called by its proper name, Viola tricolor Maxima, or by one of the many common names so frequently implied), and knowing it, does not admire the rich yet modest colors and most beautiful forms. The Pansy is an offspring of that simple little annual, heart's-ease, or violet, but the wonderful improvement made by care in hybridizing and skillful culture has left so little resemblance, that at first sight there would seem to be no relation between the two, and only a botanist or those acquainted with its history, would believe that such a beatiful flower originated with so humble a parent. It is true, however; and by the efforts of intelligent cultivators, the insignificant little flower, scarcely the size of the single wild violet, and with coloring confined to dull shades of purple, yellow and white, has grown to be the magnificent Pansy of to-day, combining such rich variety of colors and shadings not to be found in any other plant in our long list of beautiful garden flowers. Few objects are really more pleasant to look upon than a good bed of Pansies, with their cheerful knowing faces looking skyward. They are the first to greet us in the Spring, and the last to bid adieu in the Fall.

The requirements in Pansy culture are few, simple and easily bestowed. They delight in cool nights and moist days, with a fairly rich soil. While it is possible to propagate the plants from cuttings, it requires considerable care and attention, and it is not a desirable plan, especially in view of the readiness with which they may be raised from seed. Plants thus raised have greatest amount of vitality, and are better qualified to withstand the effects of a trying climate. For the production of good flowers, the plants should be young, vigorous and making a rapid growth. Hence the first essential requisite to the successful cultivation of Pansies is to produce good seed. As Spring and early Autumn seems to be the most desirable seasons at Northern latitudes in which to have the plants bloom, therefore the seed should be sown from the middle of August to the middle of September, in a prepared bed of light soil, that is moderately rich and fine. Cover the seed lightly, and press the soil over them gently, afterwards watering with a fine spray. As soon as the p'ants are large enough to handle, transplant into cold frames, or to the bed where they are wanted for blooming, having a southerly exposure, (any ordinary boards may be nailed together, or held in position by stakes driven into the ground, to form a frame), and as frosty weather approaches place the sash on, leaving them drawn several inches at one end, to admit a free circulation of air; later, as the weather becomes severe, it is admissible to give additional protection of some sort. The plants thus wintered will be stocky, vigorous and well set with buds by Spring, when the covering should be removed as the frost leaves the ground. In mild Winters, blossoms can often be gathered from the beds for the window.

The Pansy being a plant that delights in moisture and good living, it is well to water them in dry seasons, and the use of liquid manure occasionally will produce a marked improvement both in size and quality of flowers, especially is this noticeable during the Summer months, when the flowers are usually small. Keep the soil well stirred, and all seed pods (if not wanted) removed, will assist materially in prolonging the flowering period. For Autumn blooming plants, seed is usually sown early in June, and the young seedlings transplanted in a bed that will be protected from the burning sun of midday.

Amaryllis.

Not many years ago this beautiful genus of bulbous flowering plants was considered beyond the successful management of the ordinary amateur, but experience of a few years, coupled with experiments in cultivating them, has convinced me of the contrary. They are very desirable plants for garden or pot culture, and to produce good results should be grown in a fibrous loam with one-fourth leaf-mould and a fair quantity of well-decayed manure. In growing them in the garden, plant the bulbs in a sunny bed or border when all danger of frost is past, with about one-half of each bulb above the soil. If convenient, or desirable, they may be somewhat advanced in growth before planting out time, by starting them in pots under glass during March and April. After planting, water thoroughly, then wait until growth appears, unless the soil should dry out rapidly, when another watering would be necessary and should be judiciously increased as growth advances.

When grown as pot plants for the window or veranda, the requirements are virtually the same as if planted out, except that the pots should contain a good supply of drainage material, and in the treatment of the evergreen varieties, which usually retain their foliage all the season, and therefore should not entirely dry up. After blooming, much of the future usefulness of a bulb depends on a strong, healthy growth to mature it before the leaves cease activity, and they should have just as much attention at this period as when first started, if you would have them do well the following season. From the time growth commences it advances steadily until the blooming period is past, then after a few weeks the foliage will begin to ripen, indicating a period of rest, (a process that is quite essential with all flowering plants), and the supply of water should be gradually diminished. While resting they will require but little space, and may be kept in any cool, dry location.

Seedlings and off-sets should be kept growing till they are of a blooming size, or at least till they are quite large, and then receive the same treatment as others of the same variety.

There are many fine standard kinds of both Spring and Fall-blooming varieties, of which a good selection is here given. I would first add, however, that when desired for Fall or early Winter blooming in the window, the bulbs should be repotted in the Spring and after watering well, turned on one side in a shady place for the Summer resting period. Examine occasionally to see that the soil does not dry out entirely, and toward Fall encourage a vigorous growth by setting the pot in a warm sunny position and supplying water as required. A. Johnsonii is the finest variety of the Amaryllis family. It is a stately plant when well grown, and usually blooms twice a year, the flowers coming soon after the leaves. The leaves are a dark, rich green, linear shaped and quite long. A. Valotta Purpurea is quite similar to Johnsonii in habit, and both make fine pot plants, requiring a like treatment. After blooming, the water supply should be gradually diminished and the bulbs allowed to ripen and rest, but being of an evergreen nature, the foliage may attain its full size for several weeks. A. Belladonna is a very fine variety and has the largest bulb of any. A. Formosissima, (Jacobean Lily as it is sometimes called), is a unique variety with velvety, scarlet flowers that consist of six petals, three erect and three lower, forming a curious shaped blossom. These bulbs are usually planted in the open ground in the Spring and wintered in a cool dry place, but they may also be grown in pots like hyacinths with good success, if kept dormant during the Summer. A. Vittate Hybrids, and Equestris are among the new varieties of considerable merit, while A. Tretea (Fairy Lily) is a native of Florida of recent introduction. The flowers are very fragrant and pure white.

Begonias.

Few plants are more interesting or more admired than the beautiful Begonias. Their beauty of foliage, combined with graceful flowers, handsome colors and free blooming qualities, tend to make them most desirable plants that grace our windows, conservatories and gardens.

Begonias are natives of tropical countries in both the eastern and western continent, where they inhabit the mountainous regions at a considerable elevation. They belong to the natural order Begoniaceæ, and carry their generic name in honor of Michel Begon, a naval officer, who first brought them to notice about two hundred years ago. Since their introduction, a rapid advance has been made in the production of hybrid varieties, which combine most valuable qualities of the best sorts. Every season brings out new members of this aristocratic family, and the improvements which have been made within the past few years, are almost wonderful. The greatest variety exists in the peculiar forms, sizes and surfaces of the leaves, which, upon most kinds, are one-sided, being larger one side of the mid-rib than the other. This does not at all detract from their beauty, however, for their outlines are so graceful that the diversity is far more pleasing than mathematical symmetry. Not only in the leaves do these plants present peculiar features, but also in their flowers, which are both staminate and pistillate in each cluster. Their position on the outer edges of the branches of some. The semi-pendulous habit of others, and the brilliancy, grandeur and grace, as well as the style of character, great distinctiveness and freshness of all, are characteristic qualities seldom combined in the same plant.

The Begonia, like the geranium, is very seldom attacked by insects, but it must be provided with good drainage, or trouble from sour soil will follow, causing much disappointment and often ruin to the plant. It is not as strong and hardy as the geranium, and, of course, should not be expected to grow as sturdily under the conditions which that plant frequently has to contend with, but it can be easily grown, and will amply repay any one for the small amount of attention required. Like the fuchsia, the begonia does not care for much sunshine, an almost entire shade seems to suit it best. The soil should be rich, light and porous, similar to that in which fuchsias flourish well, and a liberal amount of leaf-mould is just what they seem to want. A moist atmosphere is always desirable, and much can be done in this respect by frequently syringing when the sun does not shine on them, though I would exempt most varieties of the Rex class. These succeed under glass globes or in ferneries, the air being always moist and of a uniform temperature, but pretty good results can be attained without glass coverings. The foliage should be kept as free from dust as possible by covering with a newspaper before sweeping, and also by blowing or brushing off the dust that will accumulate in spite of all precautions.

The Begonia family is divided into three distinct classes, of which flowering or upright growing varieties comprise a majority of the entire family. The other classes are divided into Rex, or ornamental leaved, grown principally for their large and beautiful foliage, and the Bulbous, or Tuberous Rooted, from the fact that they are grown from tubers similar to the Dahlia, though, of course, much smaller. Of the flowering varieties one need not make a poor choice, as all are very beautiful, and when grown as specimen plants it is well nigh impossible to equal their beauty. Some sorts, of course, combine more good qualities than others, and among the best of this class a good selection can be made from the following:

RUBRA.—If you can have only one Begonia, let it be the Rubra. Its large, showy panicles of coral-colored blossoms, stand out well above the handsome leaves of rich, dark green with fine effect. The strong, stiff cones grow rapidly and often reach six feet or more.

Grandiflora rosea is an appropriate name for a very fine plant. It is also a strong grower, and becomes a good-sized bushy plant the first season. The waxen, rosy-pink colored flowers are borne above the dark green lanceolate leaves, in drooping clusters, that contrast most charmingly.

Wiltoniensis grows in a bushy, compact form without much attention, and is especially desirable as a Summer plant for balcony or window box, where it will be somewhat sheltered from the strong sunshine. The foliage is of a bright, rich green, shaded with darker tints, while the stems and veins of the leaves are dark red. Now, combine the beautiful carmine-pink of the flowers which droop gracefully above the foliage, and you may draw a faint idea of the effect. Brannti, like Wiltoniensis, grows in a bushy, shrub form. Its deep olive-green leaves increase in size and number very rapidly, and the graceful clusters of blossoms are borne on long stems well above the foliage is a very striking contrast; it makes a handsome plant for the window or veranda.

Semperflorens gigantia rosea is a very satisfactory Winter flowering plant. It is a strong growing variety and blossoms freely about Winter.

Alba Perfecta, or Rubra Alba as it is sometimes called, makes a strong upright growth quite similar to Rubra, the flowers being white instead of coral-red. Another type of this variety is found in Argyrostigma picta, or Alba picta as some call it, having smooth glossy leaves, about the same size and shape, and of a silvery green color, spotted white. The flowers are of a lemon-white color.

Some begonias seem to belong to a separate class, between the flowering varieties and Rex, combining many characteristics of both.

Among these are Margarite, Metallica, Argyrostigma picta, and Subpeltatum Migricans.

The Rex class is without doubt the most charming of all foliage plants. They seem to feed on the rich metals of the soil and to spread the lustrous tints over their leaves, which thus become an illuminated map of the mineral kingdom. They are not adapted to window culture, because of a lack of moist temperature, and the too frequent changes of temperature, yet they can be grown there, and often with good success. Keep the soil somewhat dry, for too much water at the roots will cause an unhealthy debilitated appearance. The leaves are generally brittle and their surfaces thickly beset with hairs; making it difficult to remove any dust that may accumulate without injuring them, and it is therefore advisable to cover the plants well before sweeping. Species is an old, but meritorious variety, of a beautiful metallic lustre, with center and edge of a soft velvety green, and a broad silvery zone, terminating in the point. Louis Cretien is a fine plant of exceedingly beautiful coloring and a very high lustre like changeable silk. Marquis de Peralta, Countess Erdody and Lesondsii will form a very good collection.

The Tuberous-rooted section are among the handsomest of our Summer flowering bulbs, and although comparatively new to many of our flower lovers, its popularity is steadily increasing, and bids fair to become a strong rival of the geranium as a Summer bloomer. They are as easily grown as geraniums and have many points in their favor. A compact habit, shapely glistening foliage, crowned with an unbroken mass of handsome flowers having almost unlimited range of color, make them desirable for either pot plants or bedding. The tubers should be repotted in March, using a light rich soil and water when they seem to need it. If grown in pots during the whole season they will need shifting as growth advances. They are just the thing for window or veranda boxes in partially shady places, and produce a grand display.

Towards Fall the plants will show signs of needing rest by donning a yellow tinge to the foliage. Then reduce the water supply gradually until the branches have fallen, and by this time the soil will be quite dry, when the pots may be set away in some dry place free from frost, without disturbing the tubers, and remain there till Spring again. When the plants are bedded out, sometimes it is necessary to put them in pots or boxes, and bring them inside for completion of growth, as severe **early frosts** are likely to injure the tubers.

To those not fortunate in choice of Winter quarters for plants, the tuberous section of begonias are particularly suitable, as they can be carried over Winter, after blooming all Summer, quite as well as gladiolus, and the latter part of March, or early in April, may be started in pots, first shaking them out of the old soil. Use fresh rich soil and water as growth advances.

Lawns, from Seed.

A handsome lawn adds greatly to the attractiveness of any place and it is much cheaper to obtain it by sowing seed than by sodding, and you will also in this way avoid many noxious weeds which are in the sod.

The ground should be graded to the proper slope to secure drainage, and if but naturally rich, fertilizers should be added before sowing. For this purpose bone dust, crushed bone, or fertilizers containing bone and potash are the best. Three to five hundred pounds per acre is usually sufficient. For small plots ten pounds to about four hundred square feet.

Work the soil by plowing or spading until thoroughly pulverized, being careful to leave the entire surface as near alike as possible that the grass may be even in its growth; finish by harrowing or raking until made fine, and finally level by use of heavy roller.

Seeding may best be done in Spring or Fall. With the ground prepared as directed, let the surface be gone over with a fine rake, and the seeds be evenly scattered, after which carefully rake or brush the seed in and follow with the roller. To secure the best results use plenty of seed, four bushels per acre being about the right quantity. One pound is sufficient for six hundred square feet. Let the grass obtain a good start before cutting, say a height of three or four inches, as the growth is retarded by too early cutting. When well rooted a lawn should be trimmed with a lawn mower about once in ten days. In very dry weather a thorough wetting should be given about once a week; a little water on a dry surface often does more harm than good.

Old lawns can be improved and renewed by the application of fertilizers and seeding about one-half the quantity required for new lawns.

Oleander.

How old an oleander must be before it will blossom will depend somewhat upon how well it is treated. Some plants bloom before they are a year old, and most of them blossom the second season. To grow it well, it should have a rich, light soil, and plenty of water should be given when it is growing rapidly. It has a mass of fibrous roots, and they like a light soil through which they can easily penetrate, therefore a stiff, heavy soil is not at all to its liking. If you want a tree of it, let it grow to a height where it seems desirable to have the top begin, and then cut it off. Branches will be put out below, and these should be cut back when they have made a short growth to force them to branch, and so help in making a bushy top. I prefer to grow this plant as a shrub, because this keeps it down in the room, while a tree in time becomes so large that the top of it comes way up to the top of the window, out of good light. You will do well to put your plant in the cellar in Fall, giving it only enough water to keep it from drying up. In March, you may bring it up and give water, light and heat, and in a short time it will begin to grow, and soon you will notice buds appearing on the ends of the branches. In Summer you may keep it on the veranda. It is never advisable to put Oleanders in pots in exposed situations, for they dry out and the blossoms are never as large and fine as when in a somewhat shaded place. When growing or blossoming they must have a good deal of water. I would not advise you to turn your plant out of its pot into the open ground because the roots will spread out so far that when you come to take it up in the Fall you will have to cut off a great many and this will seriously injure the plant. I have two old plants which I keep for blooming in the open ground, but I never think of putting them in pots in Fall. I lift them and crowd their roots in a large box, which is put into the cellar and receives no more attention until April, then it is brought up and the plant soon begins to start. As soon as the weather becomes somewhat warm, I set the plant in the open ground, where it gives a profuse crop of beautiful flowers, and often two during the season. Treated thus the Oleander can be made one of the ornaments of the Summer garden and with as little trouble as it is to lay down a tender plant and cover it.

Planning the Garden.

It is Winter—nature is at rest. We saw the last rose drop its petals and now we must wait the dawn of a new Spring morn before they open again to us their eyes of beauty. While they sleep let us think and plan for the coming of Spring. To a person of refined taste it is not enough to have a well formed and nicely filled flower garden; it must be nicely arranged. Though our garden may have been to us a source of great pleasure and enjoyment, the arrangement may not have been altogether to our standard in taste, the colors did not harmonize to suit our fancy, plants that should have been in the background shown too conspicuously in front, while there may have been a stiffness in outline that destroyed much of the fine effect intended to be produced. It is of the utmost importance in laying out the flower garden that the outline of the beds in their relation to each other should be good. Then they should be filled with plants that either harmonize or contrast in height and color with each other, according to the effect intended to be produced. A good plan is to make out a list of just such plants as are desired for the different beds, with their relation to each other as to height and color, having previously drawn on a piece of paper a plat of the garden, shape of beds, etc. A practical rule is to place the tallest plants and the most intense colors in the center of the beds, using the shorter and less decided tints for outer circles and edging with very low or creeping plants like Lobelias. Beautiful flower beds that will always attract attention may be made of intense crimson or bright scarlet for the centre, then edge with a broad band of pure white and a narrow outer circle of green. White works in well anywhere, but in large beds where several colors are used, makes a much finer effect when placed between two intensely bright or very dark colors. Tall plants like Cannas look well as single specimens, or grown in clumps with a double row of scarlet Geraniums, edged with Dusty-miller for outer circles. In arranging flower beds plants should be used that come into bloom at the same time, or nearly so, though if the foliage is good this is not always essential.

Gladiolus.

Gladiolus are easily raised and certain to bloom; it is a wonder that they are not more generally cultivated. From year to year they are being improved, both the Gandavensis and Lemoinii Hybrids; some of the old named are (or ought to be) discarded and new and better added to the long lists; especially in the separate colors as red mixed, white and light mixed, etc., the slaughtering of the poorer colors and smaller blooms goes on with recklessness, as those that see it will say.

Raised from seed many will bloom the following season while others take their time, even three years before they bloom. There will be some that may not satisfy our fancy, such of course we discard, while many will be all we may desire; larger and better markings.

Once in a while we get a real surprise. Such I had six years ago. It was a flower of extraordinary size and beauty, ground color pearly white, striped with rich carmine shaded rosy purple, velvety texture, the least affected by heat or drought, very strong stiff stem and large spike; bulbs are very large. Although I used all known means to propagate it I succeeded to grow only 100 large bulbs, which the Iowa Seed Co. purchased and named Royal Queen. It forms very few bulblets.

I grow seedlings every year from seed of the best flowers and made great improvement in the quality of the flowers, yet such as the above mentioned don't grow very thick. I have again a few that I call "Beautiful Monsters," but it will take several years before they will be for sale; there are two double among them like a semi-double rose, of pink color, that bloomed the first time; next Summer will tell whether they come true. They had no bulblets. It is a pleasure to watch for the seedling's first flower to open, that none but flower-lovers can understand.

Abutilon Thompsoni Pleno.

The double flowering Abutilon, A. Thompsoni Pleno, is one

of the most beautiful of the group of plants that are popularly known as Flowering Maples. It grows from two to five feet in height, and produces beautiful bell-shaped drooping flowers continually throughout the year.

The flowers resemble in shape and form a double Hollyhock, and are of an orange color, veined and shaded with crimson, and they remain in perfection for a considerable length of time. The foliage is also very attractive, being beautifully mottled with green and gold. It can be planted in the open ground at any time after the middle of May, but if wanted for Winter blooming, should be taken up and potted early in September, so that it can become well established before cold weather sets in. For the Winter give them a light sunny situation and a temperature of from 50 to 55 degrees, and as soon as the pots become well filled with roots give liquid manure once or twice a week. In potting use porous or soft baked pots, and let them be proportionate to the size of the plants; drain them well, and use a compost of two-thirds turfy loam and one-third well-decayed manure. Pinch back the leading shoots occasionally to secure nice, compact specimens, and spray or syringe the plants occasionally, to guard against insect pests. Grown as above advised, this Abutilon will be found to be one of the most desirable of window plants, and one that should be found in every collection.

Wild Violets.

In beginning this article, I believe I cannot do better than to copy a few lines from the letter of a dear friend, an ardent lover of wild flowers, Mrs. F. E. Briggs of Washington. She writes: "There is a violet I would give much to see again, the "wood violet" of New England. A friend in the South sent me seed that she assured me was of that kind, which I carefully nursed, but it proved to be an altogether inferior kind, a straggling, weak-growing white violet, a mere shadow of my old favorite that grows a foot high, firm, yet graceful, flowers as large as pansies used to be; white, tinged with purple; blooms a long time, has a pleasant woodsy smell, though not the true violet fragrance. Only one of our native violets is truly fragrant, and it is a tiny thing with stems a little over two inches high; white, with brown lines. I never found it in but one locality." In reply, I said, you refer to the white "wood violet" of New England. From your description of it I believe that I once had the same white violet. This, to my amazement, grew over two feet high. It piled and massed up among the other plants, indeed it seemed almost inclined to climb. The flowers were white, outside of petals were tinged with light purple; they had a pleasant, sweetish or "woodsy" fragrance. They bloomed continuously from May until November. The bed in which they grew contained many wildlings. The soil, rich and deep, was from the woods. To my extreme regret I lost the violet. It, with the bed, was swept away in the great flood which devastated the Ohio Valley in February, 1884. I have never been able to replace it. But I will certainly make a search for this charming "wood violet" next summer. Indeed, I believe I will make a bed especially for wild violets. I am sure I can find four varieties of blue, or rather purple, and two of white and two of yellow, of deep golden yellow, veined brown, and if I can find a white violet like the one referred to above, I think I will have a trellis made for it, and try to coax it to climb. Do not my friends, I pray you, do not laugh with disdain, and say "who ever heard of a climbing violet." I do not know that any one ever heard of one, or ever saw one, but you may depend upon it if I can coax a violet to climb up a trellis, I shall certainly coax. And I will tell you another thing, I fancy that that violet would make an elegant basket plant—pretty and dainty, and sweet and blooming all Summer and Fall. I fully believe that if the same care and culture were given to our native wild violets that is so lavishly bestowed on pansies that the result would be equally as satisfactory and as pleasing.

A Cheap Greenhouse.

Several years ago, the cramped position which often became necessary in attending to the culture of plants in a cold-frame suggested the idea that something might be constructed at a comparatively small expense which would not only relieve the necessity of much creeping on hands and knees, but might also be made serviceable during a large portion of the year as either a greenhouse or hothouse. Acting upon this suggestion, the result was a structure seven and a-half by ten feet, internally, the sides being composed of cedar posts, boarded inside and outside with common boards, the space between being filled with sawdust. The front was two feet above the ground, while the rear was as high as the length of the posts would allow, the sides forming a regular slope from the rear to the front. The lower part of the roof was composed of ordinary three by six feet hot-bed sash, and the upper was made double of common boards, battened on top, leaving a space between the upper and lower boards of ten inches, which was also filled with sawdust. In one corner, at the rear, a wooden chimney four inches square was inserted, and about midway between the sides was placed a trap-door for ventilating purposes. Finally, double doors of rough boards, battened, were placed at the rear end of one of the sides. A pit two feet deep was next dug on the inside, of such a size as to leave a space undisturbed, three and a-half feet on each side, and three feet along the rear. On the two sides rough benches three and one-half feet wide were constructed; the one running the entire length of the side, and the other extending from the front to the doorway, being partitioned off from the latter to prevent cold draughts from the doorway striking the plants. The boards covering the benches were laid on loosely, so that they might be removed at pleasure. The purpose of the pit was to afford standing and working room, and to form a place wherein to set a coal-oil stove. The use of the stove was at first not at all satisfactory, the coal-oil odor being unbearably offensive, notwithstanding the wicks were not untrimmed by scissors, the loose charred parts being removed by means of a soft brush, and every part of the stove being kept scrupulously clean. There was incomplete combustion, and it was due to some defect in the construction of the stove (an Adams & Westlake, having two four-inch burners). In justice to these manufacturers, it should be stated that this fault does not pertain to all their stoves of similar size and pattern. But in the end this unpleasant feature proved to be a benefit, for it led to a device for the removal of the odor; so beneficial in other ways that its application to either odoriferous or odorless oil stoves is desirable. This device consisted of a small sheet-iron drum placed upon the stove, and connected near the top with ordinary three-inch water pipe, such as is used to conduct water from the roofs of buildings. The first joint of this pipe was a short one, and made to fit loosely, so that it could be easily disconnected from the drum by shoving back into the next joint when it became necessary to lift off the drum for the purpose of cleaning the wicks and stove. This pipe was then led through the bottom of a long, narrow box, placed in front of the rear bench, and kept filled with damp sand, to be used when bottom heat was required. The pipe was then carried along the rear wall, and next, upwards, until it terminated an inch or two below the open bottom of the wooden chimney; were it carried directly out through the roof, the draught would be strong enough to extinguish the flame in the stove. There should be sufficient length of pipe to absorb all the heat from the stove, and if made to run backwards and forwards this end will be better attained. If a little of the coal-oil odor does not occasionly escape into the room it may be advisable to make it do so, for the mealy bug will not thrive where it is present, and the odor in small quantities is not injurious to the plants.

By placing shelving at available places a large additional amount of plant space can be obtained. In the building described there are over one hundred plants, quite a number of them being specimen plants in eight-inch pots, and but very few in pots as small as four-inch. Yet they are by no means crowded, and are in a very flourishing condition, all of them being hot-house plants. With an outside temperature of not less than fifteen degrees, an inside temperature of from fifty to sixty degrees can be easily maintained at an expenditure of one gallon of oil for each twenty-four hours. Should this temperature of fifteen degrees be accompanied by strong winds, or should it lower to zero, some blanketing with old carpets or straw matting over the lower part of the sashes will be required. How much will depend largely upon the manner in which the glazing has been done. On sunny days the consumption of oil will be greatly lessened.

The great value of such a structure, however, is for the late Fall and early Spring use. It is best to transfer the plants some warm day in December (as warm as can be expected at that season) to Winter quarters in the conservatory or living room, and make use of the building for a different purpose. Having made this transfer, remove the boards which form the top of the benches, spade up the ground underneath, and put in low plants which require Winter protection, covering them, as the cold increases, with straw ; but, as the room will be required for warmer work in February or March, discretion must be exercised as to the nature of the plants. They should be either such as will be desirable to force into early growth, or that can, in late Winter or early Spring, be safely transferred for a few weeks to the cellar. Last Winter mine was filled with the Summer layers of perpetual carnations and tender roses ; this Winter I purpose to fill it with anemones, ranunculus, etc., all in pots, to be transferred to the cellar in February or March, and again to the open ground, turning them out of the pots as soon as the weather will permit.

Hot-beds.

All flower lovers are to some extent interested in growing flowers and plants under glass, and while some may not have a convenient place or room for a construction of this sort, others may, and no doubt would be pleased to learn something about them and how they are constructed, so we have devoted a chapter to this subject from the pen of practical persons.

Seeing various articles on how to make a hot-bed, and hoping that a few suggestions from one who has actually made and taken care of hot-beds and cold-frames, may still be in time, I offer the following for the benefit of new beginners (experienced gardeners need not read any further).

First, the pit should be eighteen inches deep, and one foot longer and wider than the frame, which should be six feet wide. Manure must be fresh, and such as has already begun to heat. When filling the pit, spread evenly, and tramp firmly; fill pit level with surface, a little higher at back than front. The frame may be of one inch, rough lumber, twelve inches at back and eight inches at front, closely fitted at corners, and well-braced at bottom and top. I say at bottom, as the earth that has to be banked around the outside will press the frame in at the bottom if not braced, and the braces on top are necessary to support the sash or sheets; after the frame is made, place it directly on the manure, but never set it on posts or brick walls. The object is to let frame, earth and manure all settle together, as the latter decays; five or six inches of rich garden soil may now be spread evenly over the manure,

Where only one hot-bed is made and all kinds of seed sown, it is necessary to put a tight partition between plants that require much heat, such as tomato, peppers and egg plants, and those that do best in a lower temperature, cabbage, cauliflower and celery.

I always use a thermometer, set about the center of the bed, and one sunk in the earth about two and one-half inches, for you want to know the temperature of the soil as well as the air.

Tomatoes, peppers, etc., do best at 70° or 75° while cabbage thrives best at from 50° or 60°

When watering a hot-bed do it thoroughly; be sure that the earth is wet down thorough. It is best to heat some water in the house and mix it with the fresh water before using. If the heat in the hot-bed gets too high, sharpen a broom handle and thrust it through the earth, well into the manure, making holes every two feet over the bed. On the other hand, should the heat go down, it can be renewed by making holes as above, and pouring in boiling hot water and closing the holes again. Muslin will do nicely, instead of glass sash, if given a coat with a paint brush of the following: One quart boiled linseed oil; two ounces rosin, well pulverized, and one ounce of sugar of lead, all dissolved in an iron or tin vessel and made quite hot.

The Finer Aquilegias.

The better forms of Aquilegias, such as Chrysantha, Coerula and Glandulosa, and the seedlings of these I class among my most attractive hardy plants of the garden. Besides being so perfectly suited to border culture, I find they do exceedingly well in pots, thus adapting them for window or veranda culture. Not the least advantage is that they can be easily raised from seed.

Let me mention my method of raising the plants; I sow the seed in the end of July or early in August, in low boxes of sandy soil, which latter are kept in a frame covered with whitewashed glass. As soon as the plants are large enough to handle they are pricked out into other pans so as to have room to grow, which they will do until late Autumn. Towards Winter, and even during the Winter, the soil should be rather dry, for, although Aquilegias like plenty of root moisture when they are making active growth, the roots will perish during the Winter if the soil about them is at all wet. Early Spring is quite soon enough to think of moving them again, which should be done into a well-prepared spot into the garden, unless they are to be grown in pots. In any case the soil should be moderately rich and deep, and the surface level, for if the weather is dry they will stand plenty of water, both while they are growing and while they are in flower.

With me the varieties of coerula will flower for three or four years, but they do not do so in all soils, more frequently dying away after they have flowered twice; therefore, it is necessary to raise a stock of young plants, if not every year, once in two years.

Those plants required to flower in pots should be taken up

and placed in six-inch pots at the end of October, and the plants kept in a cool pit all the Winter. In this structure they will come into flower in May, and will make a striking object for any position where decorative plants are required. As soon as they go out of flower the stems may be cut down and the plants turned out again into the bed whence they were taken.

Ornamental Grasses.

Hardy ornamental grasses, when grown in large masses or clumps, form most beautiful and striking objects, whether grown upon the lawn, in the border or back grounds.

All through the Summer their long, airy, graceful leaves dance about on every passing breeze, and in Fall and Winter their large, feathery plumes are fitting accompaniments to the brilliant Autumn leaves and everlasting flowers so largely used for Winter decorations.

Foremost among the many beautiful and valuable plants, introduced from Japan by Mr. Thomas Hogg, is the Euialia Japonica zebrina.

Planted in deep, rich soil, and liberally mulched each Winter with manure, it will attain the size of six by six feet, or even greater dimensions. It is perfectly hardy and increases each year in size and beauty. But the most distinctive feature of this beautiful grass is in the variegation of its foliage. The leaves are a beautiful dark green, and striped crosswise instead of longitudinally, with creamy yellow. Its beautiful lyre-like plumes are plentifully produced, and when dried are valuable for decorative purposes, especially in bouquets of dried grasses and everlastings, where the noted pampas plumes would be objectionable on account of their size.

A fitting companion for this beautiful grass is E. Japanica variegata.

We have no hesitation in saying that a well grown specimen of either of these deservedly popular grasses is an object at once striking and beautiful. Crianthus ravennaea is another very desirable grass, growing larger than either of the above, with a wealth of sea green, ribbon like leaves and large plumes, produced at the top of tall stems. It is hardy and very fine, much resembling pampas grass in general appearance.

Gynerium argenteum is the true pompas grass from South America, which produces the beautiful silvery white plumes of commerce. These plumes cannot be produced here in our northern climate, but are grown to perfection in California, and we see no reason why they may not be successfully grown at the South.

Cultivation of the Cactus.

Ladies frequently say to me, "How do you get your cacti to bloom so nicely?" One must understand the natural habits of a plant in order to cultivate it successsfully. I remember, many years ago, of seeing my father plant a cactus slip in a pot of pure charcoal. It grew very luxuriantly, and when it bloomed the flower exceeded in size and beauty any we had ever seen on the old plant. I think four things are essential in the culture of the cactus—small pots, little water, rest at the proper time and plenty of charcoal in the pots. The roots should not be disturbed very often. I have a plant six years old, growing in a quart can, that has never been repotted. It has large three-cornered leaves one and a-half to two feet long, and so heavy it will not stand alone. The flowers on it now are beauties. A large pink flowered one, sixteen years old, is growing in a-half gallon jar, has been repotted once in that time, it blooms freely twice a year. When there is danger of frost, I bring all my cacti into the house and set them in some out-of-the-way corner where they will not freeze. Never give any water until February, then give them a good soaking with warm suds from the washtub and set them in a warm place. The leaves will soon begin to freshen up, and they will usually be in bloom in about two months. After blooming they make their growth. When you think they have grown enough lay them down in some safe sunny place out of doors and let them dry up till they are well wilted; set up your plants and water as before, and you will have another crop of flowers before Winter. In choosing slips take old wilted leaves or those showing flower buds. The buds will usually grow on and bloom, and if given proper treatment will bloom the next season. The lobster or crab's claw cactus is beautiful for a hanging basket. Having kept cacti for ten years or more with never a flower until I learned to let them have their dry rest, I now have no lack of flowers.

Wallflowers.

This old favorite seems to grow and thrive in England without much care or cultivation, and for the sake of old associations I have tried to keep them here, but found the greatest trouble was to keep them cool enough in ordinary houses. They do pretty well if the seed is started in beds and transplanted to large pots, plunging the pots in the ground, and if intended for flowering in the house, they should be removed before frost and given a cool place, but they are not recommended for house culture, as it is a waste of room that could better be filled by plants that would furnish more flowers with less trouble. They may be stored in a cool place, with moderate light and enough water to keep them from drying up, and early in the Spring planted out in the open border. Under such treatment the old leaves wither and either fall off or are easily removed, when a new and vigorous growth will soon start, followed by an abundance of bloom early in the season. They do not Winter over a second season satisfactorily, so that young plants should be produced each season if flowers are wanted the following year.

An Amateur's Greenhouse.

I would like to give you an account of some very pleasant and successful experience I have had in growing flowers in Winter, by which your readers may be informed that the luxury of a hothouse may be enjoyed without the usual requirement of being "well off." I live on the north side of a street running east and west. My rear fence is ten feet in height, of tongue and groove heavy yellow pine. Against this I built a little hot-house seven feet square, front height six feet, and back nine feet. Having about my house all the necessary tools, and being accustomed from boyhood to their use, I undertook to build it myself, and made a good job of it. The house finished, I stocked it with plants for the Winter, and arranged a square tin vessel under the lower stand, filled with water, below which I placed a coal oil lamp for heating. This I found destroyed nearly all the foliage in the house. I then had constructed a tin tank, seven feet by twelve inches and four inches deep, with a hollow bottom two inches deep; into this hollow bottom I had three large holes cut for three lamps, and in the end of same a pipe was attached to run outside of the house as a chimney to carry off the fumes of the lamps. This tank filled with water required but one lamp placed under the middle hole to keep the house warm during the severest weather—the house being covered at night with carpet over the top and front, and my flowers prospered and grew splendidly. Over this tank I started, in boxes, all my Spring seed, and this service alone was worth all my trouble and expense. I will give you my memorandum of expenses, and hope it may tend to encourage some other lovers of flowers to possess themselves of one of the greatest pleasures and comforts they can have while engaged in their favorite pursuit:

2 old sash, 3¼ by 3¼ feet, for front....................	$1.00
225 feet flooring, tongue and groove, at $3.50.......	7.88
48 feet 2 by 3 scantling, at $2.75...........................	1.32
2 top sash, made to order.......................................	2.50
Staging, (work done by myself).............................	.60
Hardware..	1.50
Cord and pullies for top sash.................................	.80
Tank and lamps for heating...................................	9.70
Total...	$28.70

Tables, shelves, racks, etc., I built out of old lumber from recent fence repairs. I kept no account of oil, but think I used about a gallon per week for the one lamp, the largest made. Price paid for oil, 20 cents per gallon. Vick's Magazine.

The expense of erecting glass houses will vary somewhat, of course, according to the locality and cost of materials, carting, etc., but it is surprising to know how reasonable a small house may be constructed if one is at all handy with tools.

A Cut Flower Corner.

One who loves flowers will want a corner of the garden in which to raise plants to cut from, for more or less of them will be wanted in the house all through the season, and to give away in bouquets to friends, and there will be apt to be a demand from church and school entertainments for more flowers than one feels like cutting from that part of the garden designed for the adornment of the home grounds, and which cannot be encroached on to any great extent without detracting from its beauty. To supply these demands I would advise having a corner in which to sow all kinds of flowers. Such a corner can be made the most attractive part of the garden, because there will be such a "make-yourself-at-home-air" about it. When you go among your "show plants" in formal beds, one feels somewhat as if he were being entertained while making a call, but when you go into the "catch all" corner it seems as if one had "dropped in" on a neighbor with whom all ceremony could be dispensed with. No formality there, no putting on airs for appearance's sake, but old-fashioned hospitality of the free and easy sort, which is, after all that's said and done about it, the best kind of hospitality I know anything about. I always feel as if I were expected to put on company manners and "spruce up" a little when going into the show garden, but never that way when I visit the corner where I raise a "little of everything."

Such a happy family of plants as I had there this season. Sweet peas caught hold of the fence and pulled themselves up, up, up, until they could look over on the other side, and Phlox and Mignonette snuggled down among the Larkspur as if they were in love with each other; Marigolds and Petunias tried to see which could make the most show at one side of the corner, and Chinese Pinks and Portulaca made a great effort to outdo them on the other, and it wasn't their fault if they did not succeed. And in the center Bachelor's Buttons and Poppies and Mourning Brides, as they call the Scabiosa in a great many sections of the country, grew up in the most perfect harmony together, and did what they could toward making the world bright and cheerful, and that was no little. An old lady living near used to come very often over to see what she called my "old fashioned garden." "It's like the gardens they used to have when I was a girl," she said. "Then they grew flowers for their own sakes, but now-a-days they want them to help make a show. A show is what most persons care about, it seems to me, more than they do about the dear flowers."

I think she was right. If we grew flowers simply for the pleasure to be derived from their beauty and fragrance—and that, it seems to me, is why we ought to grow them—we would never plant them in prim and formal beds where the pattern and design is sure to be spoken of, but seldom the beauty of the flowers with which they are wrought out. It has always seemed wrong to me to make such a use of anything but bright foliaged plants. Use them for carpet and ribbon gardening where the design will be such a formidable rival of the flowers' beauty as to force it into the background, or rather, to make it secondary, and plant the flowers where they can be and will be admired because of themselves.

Some day carpet beds in which the effect is worked out with fine flowers will be thought in as bad taste as the floral designs exhibited at weddings and funerals. They are impositions on good taste, but they are "the fashion," and one might as well be out of the world as out of the fashion, they say.

One of the best new geraniums is Apple blossom. It is a soft pink, shading into white, almost exactly like the flower from which it takes its name. Most of our pink geraniums have been of a bright rosy shade, but this is of the pearly, sea shell tint, which is half way between pink and white; it is, in fact, a Pauline Lucca with the hint of a blush on its petals.

And speaking of Pauline Lucca, reminds me that I wanted to say that this is the very best white geranium I have ever grown. Most so-called white ones are not white, they are tinged with red or green and have a dirty look; but this variety is pure white, and it is, moreover, of fine shape and very free flowering, quite as good in most ways as the more popular scarlet varieties are. The best scarlet, so far as color and shape are concerned, is Rienzi. And the next best is William Cullen Bryant. Both of these varieties are very rich and brilliant in color, with a velvety texture in their petals, and a glow like that of gold dust on them when you look at them in the sunlight. The petals are so wide that they overlap each other and give us a round flower, like a pansy. The day of the old, narrow petaled geranium is about gone. Such kinds will soon be neglected by all who like to have full, circular flowers, and who does not? I can remember when the blossoms of this popular plant were made up of five narrow petals, standing out from each other in an unneighborly way, and about all the beauty there was about them was in their color, which has not been improved so much, in the scarlet and crimson sections, as has the form of the flower.

Among the doubles I fancy there are few, if any, superior to Madame Lemoine, which was one of the first double pink kinds introduced. It is a good bloomer, fine in shape and unsurpassed in color.

One of the best variegated leaved plants I have had for the last two years is the new geranium, Madame Salleroi. Its leaves are never cupped or drawn down about the edges as most of the green and white kinds are. Mountain of Snow and similar kinds do not retain their foliage long; about as soon as a leaf completes its growth it begins to turn yellow and soon drops off, the variegation being a sign of a lack of vitality in the strain; but this is not the case with Madame Sallerol ; the foliage is retained in a healthy condition quite as well as that of the green leaved kinds. It is a most useful plant to work in among others. I have several pots of it, and whenever there is a need of something to brighten up a group, and flowers are not at hand to do it, I use these geraniums, and their pretty green and white leaves are very effective. They grow in a very compact shape, forming a rounded mass of foliage over the top of the pot. I do not hesitate to pronounce it the best white edged variety we have. For a border out of doors it is far superior to the old section of white and green leaved geraniums.

Summer-Flowering Oxalis.

I doubt if there is anything in the bulb line which will give more satisfaction for the same outlay in money than the beautiful summer-flowering oxalis. The bulbs are quite small, but when planted in the open ground in May, they at once send up a fine lot of handsome foliage and flower-spikes bearing clusters of elegant flowers, some white and others pink, according to the variety. The whole height of the plant is about nine inches. When planted in a row for a border they make a thick, unbroken mass of foliage and flowers, and are one of the neatest and prettiest things that can be used for bordering. They bloom in a few weeks after planting, and will remain attractive all the Summer. When grown in beds, or masses, they are also effective. The bulbs should be taken up in the Fall and Wintered like gladiolus. They increase very rapidly, and a dozen or two planted in the Spring will yield several hundred when dug in the Fall. They are very cheap, and we advise those who have a garden to plant a good lot of them. For pots they are also very fine, but do not give as good a growth as they do in the open ground.

Sweet Violets.

Dainty and loveable, dear through history, beauty, sweetness and modesty, what garden, humble or aristocratic, has not its retired corner sacred to violets ? Who does not love to bend above a violet bed, gently touching leaves and blossoms, stirring the soil, and breathing in with their exquisite fragrance pure lessons of life ? Such a tender and loving reverence we feel for these sweet friends of ours, that weeds are never jerked up roughly, nor runners and blossoms heedlessly cut. Fingers touch them gently, lips press their fragrant petals softly, and the bosom which stirs beneath a knot of violets, needs no other badge of refinement and modesty, for the flaunting flower lover would never gather them.

The violet, on account of its easy cultivation and popularity, might be called the people's or the poor man's flower, for it needs no greenhouse heat, but loves a cool, moist atmosphere and will thrive where almost all fashionable flowers perish. During Winter those who grow them for show, to boast of their large size and perpetual bloom, banish them to seclusion, under a sash in a frame and sunny spot; but those who value its companionship and love to turn from their work to see the sweet bright faces nodding at them and sending perfumed messages, place them in their windows and enjoy them in the right way. About the best situation for violets is a plain board frame on the south side of the house, with a sash that may be tilted on bright days to admit fresh air. A box or pan set with the plants and placed in a south window in a cool room is its delight. In my Southern home, I can grow them in long, narrow boxes which fit the outside window ledges on the south side, but even here they do not grow so well as down in the bed where they grow all Summer, with a few leaves strewn over them, or, if I am particularly indulgent, covered with an old sash. If the soil is too rich the leaves will grow large and rank at the expense of blossoms. *It will not blossom in pits.*

Swanley White is sent out as a perpetual bloomer, but really blooms only twice a year, in Spring and Autumn. It is well worthy all the praise that has been heaped upon it, for a more beautiful and perfect flower never bloomed. In growth and bloom it is more vigorous and constant than its parent, Marie Louise; but for all this do not discount Marie too heavily, for there are few equal to her. This year the center of the flower was a salmon rose, a color I could not account for; you can imagine how lovely it looked, shading off into the pale lavender blue of the outside petals.

There are some very fine new varieties of violets, M'me Millet, containing among violets the first real shade of red. The ground color is a beautiful violet purple, shading toward center with carmine. It resembles Marie Louise in many respects. Miss Cleveland, single like her namesake, is rich, dark, bluish purple, the flowers borne on long stems, very free blooming and very fragrant. Victoria Regina has very large, pansy-shaped flowers, and is indeed the queen of single violets in color, size and perfume, but after successfully growing and blooming them, if it were necessary to give up all but two I would be well content with Swanley White and Marie Louise.

Dwarf Rocket Larkspur.

These are some of the most pleasing and brilliant-hued flowers among the annuals. They are varieties derived from Delphinium Ajacis, an Alpine plant, and all prefer a cool, moist soil. The plants are easily raised by sowing the seeds in the open border early in Spring, or, what is preferable, in the Fall, so that germination may ensue at the earliest opportunity in Spring. The seeds should be sown thin, scattering them broadcast over a space rather than in rows. If they stand four to six inches apart it is close enough; those that spring up closer to each other can be separated by transplanting. An occasional stirring of the soil and keeping it free from weeds is the extent of cultivation required. These plants are sometimes called Hyacinth-flowered Larkspur, on account of the long, narrow spike of flowers, in shape like the raceme of the hyacinth, a feature from which the term "Rocket" is also derived. The shades of color are numerous, among which are mentioned white, white tinged with blue, apple-blossom, buff, rose, brick-red, red lilac, dark lilac, azure, light blue, dark blue, violet and fawn. The varieties are seldom kept separate in common cultivation, the mixed colors appearing in a mass being quite as pleasing, as well as handy for cutting. The flowers are very valuable for cutting for vases.

In planting a flower garden we should never lose sight of those flowers desirable for cutting for the ornamentation of our rooms, and a good breadth of border should be specially provided for them. The shady side of a hedge is most suitable for those delighting in a cool soil or slight shade, while a full exposure best serves the sun-loving plants. A careful discrimination and selection should, therefore, be made of positions for plants of different requirements, if we rightfully expect to realize the highest results in our garden work. The beginner may probably make some mistakes in this respect, which will be corrected only by experience, still there is usually sufficient information at hand to guide one who carefully seeks to do his work in the best way. What to sow or plant, and how, and when, and where, are questions which the plant grower must never forget to ply. Timely asked and well-answered they form the key to most of the difficulties that present themselves to the young gardener.

The Calceolaria.

Those of us who can look back over a period of but thirty years can contemplate with satisfaction the great improvement made in that time in this very showy greenhouse plant. It is not to be expected that improvement can go on so rapidly in the future as it has done in the past, nor does it seem to be necessary, as the standard of perfection has almost been attained. The plants now in cultivation are of dwarf, compact habit. The heads of bloom are very large, and the flowers possess the requisite properties of good form, size, richness and diversity of coloring. The set of twelve plants which was awarded the first prize at the Crystal Palace contained the best examples of culture ever seen in London. The individual specimens were of large size, and the well formed richly colored flowers were two and one-half inches in diameter. One had flowers of a rich deep yellow, densely dotted and spotted cinnamon red; others yellow, blotched maroon; primrose, lightly spotted crimson; yellow, sparingly spotted with red; crimson-scarlet and yellow, self-colored flowers. How such handsome specimens are produced is a question many persons have asked. In the first place, a good strain of seeds must be obtained. Mr. James has, by careful selection through a long series of years, brought the calceolaria to its present state of excellence. He also has a thorough knowledge of the requirement of his plants. The seeds may be sown now or in June and July; they are of very small size, and a packet obtained from the seed shops is so minute, that a careless person, in opening the packet, has jerked all the seeds out of it, and innocently insisted that it never contained any. A five-inch pot is the right size, for an ordinary half-crown packet of seeds. The pot should be well drained and filled to within an inch of the rim with ordinary potting soil. The half inch on the top must be finished up with finely sifted sandy soil and made quite level; on this sow the seeds, and just sprinkle over them some fine sand. It is a good plan to lay a square of glass over the top to keep the soil in a moist condition, for if it should become over dry during the germination of the seeds, probably the whole of them would perish. I generally place the pot containing the seeds in a hand glass or frame, on the north side of a wall or fence, to prevent any injury from the action of the sun. When the tiny seedlings are large enough to be pricked out, a dozen of them may be planted in a three-inch pot, and when the leaves of these well cover the surface they may again be potted off, three into the same sized pot, to be again repotted with one in a pot. After this they grow very freely when the conditions are favorable to their perfect development; and those conditions are, first, good potting soil, composed of three parts good turfy loam, one part leaf-mould, one part decayed stable manure, and a little turfy peat. The plants must also be kept steadily growing in a greenhouse kept close to the glass, and shaded lightly from the mid-day sun. The plants must be repotted before they become in the least root-bound. They like ample ventilation, but if the wind is high and dry the ventilators must not be opened on that side from which the wind blows, as a high, drying wind causes the leaves to flag as if the plants were suffering from want of water. This is another thing that must not happen, because a plant that once suffers from over-dryness once will never make such a perfect specimen as if this had not taken place; but if this should occur more than once, the probability of successful results is very remote. Further, any plants that receive a check to their growth are almost sure to become infested with green-fly sooner than those that are kept in a healthy growing state. In fact, it must be noted here that no plant is more liable to be attacked by green-fly, which would render the plants worthless if not constantly destroyed by fumigating with tobacco smoke. Besides the raising of plants from seeds, they can readily be propagated from rooted offsets. These are obtained by placing the plants when they are past flowering into a cool pit or frame; some good compost may be placed over the bare stems, and the roots will speedily push out from the part of the stem nearest the leaves. Whenever these roots are formed the plants may be divided, and the small portions be repotted into three-inch or four-inch pots. At one time nearly all the plants grown in gardens were propagated in this way, but they are not so free in growth, nor do they make such handsome specimens as seedlings do.

The Wild-Garden.

In some catalogues are advertised packages of seeds for the wild-garden, made up of a mixture of annuals and perennials; the seeds are to be scattered in some suitable spot and allowed to grow as they will; some will find the spot congenial, will thrive and give a display of flowers and will also crowd out their weaker neighbors. There will be, no doubt, some pleasure in watching the growth of the unknown seeds, and tracing their development to a flowering stage, but the results will, by no means, be as satisfactory as those derived from a more carefully and systematically planted wild-garden.

There are very few grounds in which there is not a waste corner that might be made very interesting by the addition of suitable plants that would grow with little care and attention. And in estates of a few acres, especially in New England, there is very often a damp or ledgy piece of woods, a spring, a bushy corner, or stony field considered of no value for tillage, and allowed to remain in its natural state, such a spot may be made very attractive by adding new plants adapted to the soil represented.

To cover the whole list of trees, shrubs and annuals suited to such localities would make an article of too great length, so I will confine myself to suggestions in regard to hardy perennials, for from this class of plants would really come the most valuable and available subjects for the purpose named, as among them is a great variety in size, time of blooming, and color of flowers, and power of adapting themselves to surroundings.

Among hardy perennials are many very handsome flowering plants that are rather weedy for the flower-garden; they are weedy because they are strong, vigorous growers, and this is an objection to their use among more delicate growers, but this fault for the flower-garden is a virtue for the wild-garden, as such plants are there able to take care of themselves among the surrounding vegetation.

We have also in our gardens many plants that are out of their native condition; they keep within bounds, blossom nicely and are well behaved members of the garden family, but give them their natural home, perhaps along the edge of a brook, or in a peaty bog, they will spring into a luxuriance of foliage and flowers unthought of in their garden quarters, and, instead of a clump of a plant with its flowers you will have beds of plants and sheets and masses of flowers.

We must, in the wild-garden, work for bold and striking effects, great masses of plants and flowers that we may admire at a distance, where the roughness and unfinish will not be apparent.

Those plants that are suited to the wild-garden of large extent are equally as well suited to the corners and rough spots in small grounds, where the same conditions exist.

When a selection of plants is made for a certain spot, you must first find if they will grow there, and not only grow, but thrive and hold their own with all their neighbors. Keep in mind it is to be a wild-garden, where everything is to take care of itself and fight its own battles with but little assistance. But care must also be taken not to introduce plants that are too weedy, so much so that nothing can grow near them. A balance of power is desirable, and especially so in soil of a uniform character. In some localities a great variety in the soil would serve to limit the growth of plants to the spots best suited to their wants.

Any list of plants made up for a certain locality would not be complete for another, for conditions vary so, yet there are certain plants that should be considered in making up every wild-garden, and I will mention some of them.

One of the most brilliant of our native flowers is the butterfly weed, asclepias tuberosa. It grows on poor gravelly or sandy soil, and covers, in some localities, acres of ground. The flowers are brilliant orange, freely produced for a long season. A plant is from two to three feet high, and often spreads as much, and a poor side hill or field could be made gorgeous with masses of this plant here and there.

The Adam's needle, yucca filamentosa, a well-known garden plant, would thrive in poor soil, and groups and masses would form a striking feature in the landscape. The sword-shaped, evergreen leaves would form bold masses of color in Winter or Summer, and a more beautiful display could not be produced than that made where the immense compound flower spikes are in bloom. It will do well, too, in a rocky soil or on ledges in pockets where there is a depth of soil, and nothing could be more picturesque than to see a mass of these plants against the sky on the top of a bank or ledge, or in a wide crevice on its side.

On dry, poor soil the pretty spurge, euphorbia corollata, is very useful. It is low, one to two feet, and has great heads of pure white flower-like involucres, produced all Summer.

Many showy compositæ will succeed on most soils, and make a great display with their bright flowers. The perennial sunflower, helianthus decapetalus, and the Maximilian variety of hibiscus missouricus are very desirable.

The yellow and brown cone-flower, rudbeckia hirta, now so common in many of our fields, and the white weed, leucanthemum vulgaris, are pretty, but too common.

In good spots of soil on dry ledges several dwarf phloxes may be made to grow; the starry phlox, phlox stellaria, the moss pink, phlox subulata, and its varieties, and the rock cress, arabis alpina, and in narrow crevices and little niches the house-leeks or sempervivums, sedums, and prickly pears or opuntias, will grow. In the sempervivums there is a great variety in the rosettes of foliage, and they form elegant and interesting bunches of plants. Many of the sedums have very pretty flowers in shades of yellow, red and purple. These succulent plants will grow where nothing else will, in the dryest spots wherever their roots can find a foothold in soil or in a crack in the face of a rock. Do not be content with only one or two kinds; there is a great variety in both house-leeks and sedums, and, as they will fill a place unoccupied by other plants, there should be as great a variety as possible. In a good soil, no matter how rocky it is, there are a great many plants that may be grown; the rocks will aid in retaining the moisture at the roots. As a rule, plants with thick, fleshy roots will do well in such localities; among them are the pæony, campanula grandiflora, and the dictamnus fraxinella, and among bulbous plants the tiger lily, and the stronger thumbergianum varieties. This would be a fine place for the clematis. Imagine a rock of bush covered with C. Jackmani or other large-flowered kind in blossom; and the native Virgin's bower is very showy in such a spot.

On the edges of open, wet land the wild senna, cassia marilandica, can be planted with advantage; it forms large masses, and has a splendid show of bright yellow flowers late in the season. The mallows may be grown in wet soil. The gorgeous marsh mallow, hibiscus moscheutos, with its great pink flowers, and its cream colored companion, hibiscus flavescens, both grow in masses, and are from three to four feet high. Hibiscus militaris grows taller and is a handsomer plant, but the flowers are not so large, but even they are three inches across.

The purple and white thoroughworts form very effective groups when in flower, and the cardinal flower on the edges of streams or shady pools is a most brilliant plant in August.

In the woods the beautiful trilliums can be introduced and jack in the pulpit, violets, pyrolas and eupatoriums.

The list might be extended indefinitely to include hundreds of plants that might be made to succeed in the right positions, but enough have been given to suggest what may be used. The best lesson, however, is experience, but it is well to avoid one experience; that is, introducing very weedy plants; be careful in the selection of plants, and it is not best to try to crowd too many different kinds into one spot, but rather many of one desirable kind—enough in one mass to make a conspicuous show when in flower. Of course, in making a selection it is best to secure a succession of bloom throughout the season.

A well filled wild-garden will afford a very great amount of pleasure; we all love the woods and fields, and the flowers we find scattered in one place and another over them. Nature has a large field to work on, but often she concentrates her energies on one spot to make it more than usually attractive. In the wild garden we may assist her in her work; and if we assist judiciously our reward will be great.

Agapanthus Umbellatus.

This is an evergreen; and one makes a mistake taking it into a dark cellar to spend the Winter. The leaves call for the constant stimulus of light, and when the sunshine is withheld from the plant for the third part of the year, the leaves turn pale, sickly, feeble and flabby, and suffer not a little. It takes all the Spring and a large part of the Summer to recover from such treatment. I gave my plant six weeks of cellar confinement last Winter, however, without detriment; for the leaves did not pale, and a most vigorous bloom set in, lasting as long a period in glory as it had spent in the cellar.

The leaves of this plant are about an inch wide and a foot long when mature. They are all radical, but evidently alternate; for as one comes from the center it arches to the right, while its successor immediately arches to the left.

The leaf is entire and very plain, with only a midrib and ten parallel ribs or veins on each side of it, and it continues in vigor for months without showing any sign of decay. About the first of June, the flowerbud begins to appear amid the leaves, so flat at first as to seem only another leaf. In a week its character is fully shown, however, and then one watches day by day for the enlargement and maturity of the sheath, and for its opening to disclose the flowers within. About the first of July the sheath opens and more than fifty flower buds in various stages of development are revealed. One is ready to open in a few days, the smallest will not open for a month, and during the next six or seven weeks there stands the tall scape, surmounted by its umbel of azure, or cerulean flowers, each one nearly two inches long, on its own peduncle of equal length, one or two or even three new flowers opening every day and remaining in beauty for a week and then dropping from its support, till all have bloomed. I know not how much longer the scape would remain in vigor, surmounted with the peduncles, not without some beauty even thus. But, after waiting till the seventy-fifth flower had bloomed, withered and dropped, and two weeks longer, I cut off the scape, that all the vigor of the plant might be used in the development of two incipient plants growing all this time one on each side of the scape, but both inside of the leaves of the parent plant, each has three or four leaves from two to eight inches in length, and I am wondering whether by careful treatment I can make both bloom next Summer. The whole plant, at the surface of the ground, has a circumference of almost five inches, putting the tape outside the leaves of last Spring and embracing the butt of the flower-scape and the two incipient growths. From the day the flower-bud appeared, until the scape was cut off, my plant, in a twelve inch flower pot, stood in a basin full of water exposed to the full blaze of the sun; twice while blooming, it also received some liquid manure.

Floral Brevities.

The Germans cultivate ivy in their rooms with great success. Placing a root in a large pot by one side of a window, they will train it as it grows until it forms a pretty frame for the entire window. At Fordham there is a drug store in which ivy has been trained completely across the ceiling, passing both windows. The root from which it originated was brought from Westminister Abbey to this country several years ago.

A pretty corner is easily made with the help of a carpenter. Corner shelves may be fitted into either side opposite the entrance, and serve to hold an ornamental pot with creeping plants or a bowl with gold-fish. Such a niche, if prettily draped, could be a very great help in brightening up the hallway, which is apt, in small houses, to be gloomy and depressing in effect.

Antirrhinum.—If no seeds are allowed to form during the Summer the plants will bloom the finer, and besides throw up young vigorous shoots, making thrifty plants by Autumn, which will safely endure the Winter. A light covering should be given at the north where the weather is severe. We should not forget that profuse flowering exhausts the plants.

Scilla Siberica.—This beautiful little bulb is one of the most charming of Spring flowers. Plant them on the lawn among the grass, and they will bloom very early every Spring, before the grass starts.

Lilium Caudidum is a most beautiful and fragrant sort, and when it requires transplanting, it should be done only in the Fall, and the earlier the better. August is probably the best month to move the bulbs.

Iris.

Among the very best June flowers is the Iris family, with its large range of beautiful forms and colors. Persica is one of the first to bloom, commencing with the Crocus. Following comes, early in May, the Hispanica and Anglica sorts and the rare and beautiful Susiana. This is really one of the most beautiful of all flowers. The German sorts come later, with a great variety of color, but the most valuable class is the kaempferi. The flowers are very large and borne in great profusion, while the range of colors is very great. They are perfectly hardy and a small plant will, in a year or two, form a large clump which will make a magnificent show each year. We advise all lovers of flowers to plant a good variety of hardy Iris from the early to the late flowering sorts. They will afford a great deal of satisfaction. There is no hardy border plant except the lily which is, in our estimation, so desirable.

The Chinese Primrose.

As the time is fast approaching when seed should be sown by those who contemplate growing Chinese Primroses for the coming Winter, perhaps a few words about this most desirable house plant may not be amiss.

If you wish for an abundance of Winter flowers, do not fail to sow one or two packets of Primrose seed this Spring, thereby laying a foundation for many pleasant hours during "the long and dreary Winter," when you can count your blossoms by the hundred, instead of hunting diligently all over your window in the forlorn hope of discovering an adventurous flower somewhere, and finding "nothing but leaves."

Unless one purchases a packet of each variety of seed it is well to confine experiments to mixed seeds alone, for they are, as a rule, very satisfactory, producing so many and such diverse varieties, each of which has a charm pertaining only to itself, some peculiarity of color or making, or, perhaps a difference of form or tint in the foliage, enhancing its beauty and effectually preventing monotony, which may exist even in the floral kingdom. The seed, if sown in March or April, the young plants pricked out into small pots as soon as they have put forth a few leaves, and transferred again after an interval of a few weeks into jars of larger size, should make strong specimens in Autumn, and be ready for the Winter's campaign, especially if grown in a good strong light—not sunshine—which produces stocky plants. The chief desire of a Primrose's life seems to be, judging from appearances, to crawl out of the jar in which it is growing. This can be easily remedied by changing the abode of the delinquent, taking it out of the pot which it is trying to get away from and putting it into a deeper receptacle, removing, if necessary, a portion of the old earth from the roots, in order to admit of their being set so far down into the new jar that earth may be filled in until it reaches the base of the leaf-stalks and yet leave the requisite vacancy at the top of the pot for water.

If there is one thing that is disagreeable in watering plants it is to find one in a pot that is full of earth to the very brim, for in this case you must either go through the tedious process of putting on the water a few drops at a time, or, if tired out by this lengthy ceremony, you get reckless and pour on the water as you would under ordinary circumstances, you have the sublime satisfaction of seeing not only the water, but also a good share of the soil in the pot, make a wild rush over the side into the most inconvenient and undesirable place to be found on such short notice, no matter whether it be the leaves and flowers of some plant on a lower shelf, or your best carpet, or your last new book if you have been careless enough to lay it down within reaching distance of the deluge.

It is well to shift the Primroses into pots of larger size as soon as they have filled the smaller ones with roots, but when they have attained the dignity of a five or six-inch pot they may be placed in the window where they are to remain through the Winter, and allowed to grow on unmolested, as they will do nicely in a pot of that size.

There are two facts which those who hope for success with the Primrose should bear in mind: first, that a mellow soil or compost is an absolute necessity, from the reason that the roots of this plant consist of a mass of delicate fibres unable to contend with clay or gravel, and unfitted by nature to wrestle with hard-pan; secondly, that too much water will cause decay; a reasonable amount is, of course, required, if the best results are to be attained, but this is to be applied at the root and not on the foliage; it is better to err by giving too little than the reverse, as the Primrose will live and produce flowers with a very slight amount of moisture, while a superabundance is fatal. Like some other plants with hirsute leaves, the Primrose seems impatient of water on its foliage; it is, therefore, best, as much as possible, to guard the plants from dust during sweeping seasons on the principle that "an ounce of preventive is worth a pound of cure."

I feel rather like a culprit, an individual who forswears his allegiance to an old and tried friend, when I say that, on the whole, the Primrose is a better Winter house plant than the geranium, especially for those who have limited space. The geranium must have room and sunshine in order to grow in such a manner as to give many flowers, while the Chinese Primrose will send forth its masses of bloom in the despised north window without even the benediction of a ray of sunshine. Being a plant of lower and more compact growth than the geranium, the shelves which contain it may be quite close together. A stand with the shelves not more than six inches apart, and well filled with these plants can be made to assume the appearance of a bank of primroses, the pots and shelves being almost hidden with the foliage and flowers. Its long period of blooming renders the plant much more desirable than many another of, perhaps, more striking but evanescent beauty. The individual blossoms remain perfect for days, and as one whorl begins to fade another rises to the occasion, supplying the deficiency, and before the successive tiers on the first stalk have finished their display, new clusters are peeping up from the base of the plant to increase and intensify its beauty.

Perhaps, however, one of the greatest charms of this pretty flower is the delicate odor, so suggestive of May blossoms, and the "green things growing," that one can forget the bleak landscape outside, with its leafless trees, barren fields, or dazzling glare of snow, and in fancy step across the intervening months of cold and discomfort into the fairyland of Spring.

Fern Culture.

Many rooms which have not the light necessary for success with flowering plants during Winter are well adapted for the culture of what are termed fine foliaged plants, such as are grown for their ornamental foliage and fine habits. A partially-shaded window is just what some of our finest species of ferns delight in, and when mixed or associated with hyacinths, tulips, and other bulbous plants, a charming effect can be attained. Take a plant or two of any of the Maidenhair ferns, place alongside of them a few plants of different colored hyacinths, and nothing can surpass for modest beauty a window so filled, especially if nature is clothed out of doors in her Winter garb. Such little additions to home adornment make us feel more happy, more contented, and add to every inmate's comfort.

Ferns are easily cultivated if a few practical details are observed. Growing in their native habitats they are, for the most part, found in shady positions, where, during their growing period, they have an abundance of moisture at their roots; therefore, under cultivation, a shady window is for most kinds more suitable than a sunny one, and during their season of growth a good supply of water at the roots is demanded. While it is necessary for their success to have an abundance of water, they are at the same time very impatient of a stagnant soil, and to prevent anything of the kind occurring, perfect drainage is indispensable. Not only is drainage a necessity in the cultivation of ferns, but it is also needed in the culture of all kinds of window and greenhouse plants after they have attained a certain size. No plants do I know, except aquatics, that succeed in a soil from which the water does not pass off freely. Plants growing in pots six inches in diameter and over should have good drainage. This may be done by placing over the hole in the bottom of the pot a piece of broken pot, over this place more of the same material in small pieces, instead of this pieces of charcoal answer well. Fill about one-fourth of the pot in this manner, and over the top place some moss or other rough material to prevent the sail with mixing the drainage, and thereby preventing the water from passing freely off.

The most suitable soil for ferns is a mixture of garden loam and the black soil found in the woods, about equal parts of each, then with a good sprinkling of sharp sand through the whole, giving more if the loam is clayey and less if sandy.

Never use too large pots for ferns, especially the finer growing kinds. After potting give a good thorough watering, and keep shaded for a few days until root growth commences, after which they can be inured to the light. If possible, never repot ferns until they have commenced to grow. I have often seen valuable specimens lost by repotting when at rest. Ferns generally are not very liable to insects, the most troublesome being the brown scale, thrips, and occasionally on the young shoots green fly. The only way of getting rid of the scale is by sponging with clean water, care being taken not to injure the fronds. Old fronds when badly infested with this pest should be cut off. Nothing mars the beauty of these plants more than old and partially decayed fronds, and they should, whether covered with insects or not, be removed as soon as they appear unslightly. Fumigating with tobacco smoke destroys the thrips and green fly, care being taken not to have it too strong, as there is a risk of hurting tender fronds. In greenhouses snails are sometimes injurious to Adiantums; they eat the fronds as they start into growth. By laying pieces of cut potatoes or turnips about, the snails will leave the plants for juicy vegetables, and can then be caught and destroyed.

Some ferns are well adapted for basket culture. The best kinds of baskets for this purpose being such as are made of wire. Nothing looks much prettier than a basket suspended from the center of the window filled with a good healthy plant of any suitable fern, such, for instance as the Nephrolepis exaltata, an evergreen fern, having long, sword-like fronds. This plant looks best in a basket alone, and soon forms a large round ball, the roofs coming through the bottom of the basket, and from these are produced new plants, and then eppears a large, compact-mass of leaves above and below.

Platycerium, of Stag Horn Fern, is also a good basket plant, and, like the above, withstands a dry atmosphere for a long time with impunity.

Adiantums, or Maidenhair Ferns, make most beautiful pot plants, most of the species being well adapted for culture in the house. Their fronds are also very useful for bouquet-making, and, in fact, floral work of any kind.

Adiantum cuneatum. This specie is one of the best known, of graceful habit, and one of the easiest grown.

A. decorum. Another species well adapted for house culture, the fronds in a young state being of a beautiful pink color, becoming bright green with age.

A. Farleyence. The most magnificent of all the Maidenhair ferns; the fronds are of a pendulous habit, pinnæ large and deeply fringed, giving it a striking appearance. I have tested its qualities as a house plant, and find it one of the best for this purpose, enduring well the dry atmosphere of the sitting room.

Pteris tremular is a large-growing fern, well adapted for pot or basket; of easy culture and a bright green color.

The different varieties of Pteris serrulata are all suitable for the window garden; some of them have the ends of the fronds crested, giving them a unique but attractive appearance.

Several of the Davallias make fine basket and pot specimens, one of the most beautiful being D. Tyermanii, having long rhizomes from which are produced large, triangular-shaped fronds, finely cut, and of a dark green color.

There are several species of tree ferns which, in their young state, are very suitable for growing in the house, but soon outgrow any small space. The best of them being alsophila austrails and dicksonia antarcticar

Insect Pests.

A few brief notes in regard to the destruction of some of the most prevalent insect pests may prove useful, and their destruction should be accomplished promptly.

The green fly on house plants and roses in the open ground can be destroyed by syringing with weak tobacco water—a mixture in hot water of soft soap and a small quantity of tobacco juice is excellent. In the case of house plants the green fly can be destroyed easily by smoking them with tobacco smoke, the plants being enclosed in a box or small room. Or the plants can be dipped or imersed in weak tobacco water. After either of these operations syringe the plants with clear water.

The rose bug can be destroyed by an application of insect powder in water, or by dusting the powder over the plants and insects, with a bellows, such as are used for this purpose. The same remedy is efficient in the destruction of the common cabbage worm.

Complaint is often made of a little black beetle that destroys the flowers of the Aster. Pyrethrum or insect powder will destroy it.

Cut worms are often destructive in some fields during the month: one of the best ways to manage them is to mix one part of Paris green or London purple with twenty parts of flour, and dust it over some young cabbage leaves, and place these along through the field they infest, turning the powdered side down.

The grub of the May beetle is a very destructive creature, and it is almost impossible to destroy it directly. The best course to pursue is to make bonfires, burning old brush at night during the month, most of the insects will be destroyed by flying into the fires.

White Hellebore is the best sure destructive agent to the Currant and Gooseberry worm.

Mealy bug can be destroyed by use of a kerosene emulsion; an easy way of preparing it is to churn up equal quantities of kerosene and sour milk, and this can be well done with a egg beater. Then use a teaspoonful of the mixture to a pailful of water. The same preparation can also be used successfully against Cabbage and Currant worms; it has also been employed satisfactorily for destruction of the Squash bug. For this insect it needs to be thrown upon the plants with considerable force, such as can be given with a hand force pump; when merely sprinkled on it fails to reach the under sides of these insects, where it affects them.

We advise the use of this kerosene mixture for the riddance of any new insect pests, the means of whose destruction is unknown.

This substance is so valuable an insect destroyer and should be so well known that we shall probably be doing service to many of our new readers by here repeating some directions given concerning it last year.

"One mixture is as follows: Boil a quart of soft soap with two gallons of sour milk, and when cool add one gallon of kerosene; the whole is then churned for half an hour or more until well mixed. When used dilute with twenty times its bulk of water. Professor TRELEASE, of the Wisconsin Experiment Station, says: 'As the result of numerous experiments, I would recommend an emulsion consisted of refined kerosene two parts; fresh, or preferably sour, cow's milk one part.' The oil and milk are churned together from fifteen to forty-five minutes, varying with a temperature. The chunning requires to be more violent than can be effected with an ordinary butter churn, and the aquapult force pump is recommended for the purpose. 'The pump should be inserted in a pail or tub containing the liquids, which are then forced into union by continuous pumping back into the same receptacles through the flexible hose and spray noxxle.' When this process is carried far enough 'the liquid finally curdles, and suddenly thickens, to form a white glistening butter, perfectly homogeneous in texture, and stable.' This butter should be put up so as not to be exposed to the air, and can be used as needed by diluting it with water, two gollons to a pint of the butter."

The slugs that eat the leaves of Rose-bushes and Cherry and Pear trees, can be killed by syringing them with whale oil soap and water.

The codlin moth can be destroyed by means of Paris green and water—one poumd of the poissn to one hundred gallons of water—applied with a force pump, throwing a spray over the whole tree while the fruit is about the size of marbles, or within two or three weeks after the fall of the blossom. If heavy showers should occur, a second spraying might be necessary.

Some of these insects and many others are caught by keeping shallow vessels of sweetened water and a little vinegar added, hanging about the orchad until July, and again in the month of August.

The Apple tree and Peach tree borer, when they have worked into trees must be worked out or killed with a stout wire, and afterwards, as protection, coat the bark of the tree about the base with lye, or soap and water, soft soap and corbolic soap being best.

A Variegated Fern.

Under the popular but rather indefinite name of Silver Fern we have the well-known Pteris argyrea, one of the most ornamental ferns in cultivation at the present time. It is an evergreen greenhouse plant of robust growth, the fronds being from two to four feet in length, the pinnæ being pinnatifid and the lower pair bipartite and of a beautiful and distinct silvery white color, margined with bright green, the decided contrast in color making it one of our most valuable ornamental plants for greenhouse and conservatory decoration, and, besides, it is equally at home in the window garden. It is a plant that can be easily cultivated, and should be given a compost of turfy loam and one part of well-decayed leaf-mold, well mixed. In potting use porous or soft-baked pots, and let them be proportionate to the size of the plant, and see to it that they are well drained ; for although the plant requires a liberal supply of water, yet it dislikes to have water standing around its roots. During the plant's season of growth, which is principally during the Summer months, it should be given a warm, moist atmosphere, and a liberal supply of water both overhead and at the roots, but on the approach of cold weather the supply of water should be gradually reduced. During the Winter season the plants should be given a temperature of from 50° to 55°, and enough water at the roots to prevent them from becoming dry.

Propagation is effected by a careful division of the plant, and also by spores, the former being for amateur cultivators the most preferable method of increasing their stock. In dividing the plants, select those that have compound crowns, and cut them apart with a sharp knife, being careful to have some of the rootstock with a portion of the roots attached to each. Now pot them, using as small pots as possible, water thoroughly and place in some cool, damp, shady situation until they commence to root, when they can be removed to their former situation. In order to obtain good specimens the young plants should be repotted as often as necessary, and every available means employed to secure a rapid and uninterrupted growth.

This Pteris is rather subject to the scale, especially when grown in a dark situation or crowded among other plants. As prevention is better than any remedy, all such conditions should be carefully avoided. It is also advisable to carefully examine the plants occasionally, and if any scale are noticed, they should be carefully removed. Slugs are also very partial to these plants and one will ruin a large specimen in the course of a few nights. The very instant their visits are noticed search for and destroy the intruders at once. This Pteris is one of the best ferns we have for cultivation in the window garden, and can be grown with excellent results by following carefully the directions here given.

Window Boxes.

Window boxes for growing flowers are becoming yearly more popular, especially in crowded districts where there is no place for flower beds. Not alone in these places are they, however, beginning to be seen and used for this branch of window gardening, but for the decoration of the spacious mansion they are also well adapted, and fill a place in the floral decoration of the house nothing else can. On the upper stories this branch of gardening can be carried on as successfully as in the basement of the building. A love for the beautiful as exhibited in the culture of flowers is not confined to those having ample means and plenty of room for carrying out their desires in the way of large flower beds in Summer and greenhouses in Winter, but the occupant of the garret rooms, with no other means of cultivating a few plants but a small, dingy window, may also possess a true love for Dame Flora, and one of the best means by which a person so situated can satisfy his desires in this respect is by the use of a window box.

Window boxes can be made of a very ornamental character, or they can be made plain, but require to be of a size suitable for the space they are to occupy. They may be made of strong wire and lined with moss to keep the soil from dropping out, or they may be made of wood and decorated to suit the tastes of the owners. One main feature in window boxes is to provide ample means for the water to pass off freely. Plants will not long retain a healthy appearance if the soil gets saturated from imperfect drainage. The best kind of bottom for such boxes is made of narrow strips of wood nailed on, leaving a space of about half an inch wide between them—this allows a free passage of the water. Another good means by which the water can pass off freely, especially if the boxes are to rest solid on the window sill, is by making several auger holes along the bottom of the sides. It is not best, however, to let the box rest solid on the sill; raise it up an inch or so, or, what is better, support it on brackets independent of the sill.

The most suitable soil for filling the box is what can be obtained from a mixture of rotten sod manure; the manure should be sufficiently decayed so that it can mix with the soil freely. Place over the bottom pieces of charcoal, broken crockery-ware, or similar material for drainage, then fill up with soil, and it is ready for seeds or plants. A leading feature to be observed in filling window boxes is to have a sufficient quantity of good showy, healthy growing vines, without which a window box is a rather tame looking object. There are some vines which are strong growing, and make rapid growth when allowed a support, but are comparatively useless when allowed to droop over the edge of a box. This class of them should be carefully avoided and such kinds chosen as grow well in a drooping form. Some of the best for this purpose are here noticed.

Othonna crassifolia, a beautiful drooping plant, having dark green, succulent leaves; the flowers are small, yellow, and produced in the greatest abundance. For enduring the bright sun and dry weather we can have nothing to surpass it.

Maurandya. Of this vine there are several colors—blue,

pink, and white. It is of very free growth, producing its flowers, which are bell-shaped, in great abundance. This plant can be either produced from cuttings or seeds.

Thunbergias have considerable variety in the color of their flowers—the different shades of yellow and orange, also pure white. They are from one to two inches in diameter, bell-shaped, and freely produced. They are liable to the attack of red spider if not kept carefully watered. The plants are mostly raised from seed.

Nierembergias. Although this class of plants are generally grown for edgings of beds, and from their general erect growing habit, pinched back into bushy form, are well adapted for this purpose, still, if allowed to grow freely they assume a drooping form, and make excellent plants for boxes, especially N. gracilis, which has smaller flowers than N. fructescens, but are produced in such abundance as to make one mass of bloom.

Tropæolums are also suitable where a strong growing vine is required, but the double varieties, which do not make such a strong growth, are well adapted for this purpose. They are always in bloom, always attractive, under even the most adverse circumstances.

Besides these, there are lobelias in variety, Sweet Alyssum, the single or double flowering, or the variegated leaved double flowering variety, which is a fine plant for this purpose. English and German ivies, vincas, both Harrisonii and variegata, are suitable, the main feature being to secure good healthy plants, and induce at all times a showy, healthy growth.

For center plants, anything suitable for the flower garden can be used. When ample means can be employed, the finer kinds of foliaged plants can be used to advantage, such as the dwarf-growing palms, dracænas, crotons, and the fine-leaved caladiums. Some of the best flowering plants are begonias, especially Begonia rubra and Begonia metallica; also the tuberous-rooted varieties are well adapted for this purpose. Impatiens Sultani has no equal for this purpose; it is one of those plants which is always in bloom from the time it is placed in a two-inch pot until the frost cuts it off in the Fall. For shady windows there are fuchsias, ferns, and other shade loving plants.

Keep off all insects and decaying leaves, give sufficient water at all times, sprinkling and washing the foliage when necessary to keep it clean and healthy. Plants half cared for will soon grow to be distasteful, while the bestowal of the attention necessary to keep them healthy will ensure a reward of keen enjoyment.

Hardy Perennials for Beds.

Every progressive gardener having in his charge puolic grounds is seeking for something new to attract attention, elaborate and laborious beds are worked out in succulents, and other bedding-out plants, varying only from one year to another in design; the same plants year after year are familiar to all, by appearance, if not by name, many with no beauty except as they form a patch of color in a patch-work design. It is late in the season before an attractive display of bedding out can be made, and, as the first frosts kill everything tender, the beds are often anything but attractive after this.

Those in charge of cemeteries find an especial need for something to brighten up their grounds on Decoration Day, the one day in the year when the largest number of visitors are present. Of course it is out of the question to have a display of tender bedding-out plants as early as this. Tulips are past, daisies and pansies are nice, but they cannot be used everywhere, and in most places there are not the facilities, time or funds to propagate the large number of plants that would be required to fill out the beds, and, in fact, this is the obstacle in the way of many places, public and private. What is wanted is something not expensive, easily propagated, and easily cared for with little labor and expense—something we can always depend upon for a display at the desired season.

Among the hardy herbaceous plants are several that will fill the bill and be of the greatest value for Spring and early Summer beds. The varieties of phlox subulata, the moss pink, will stand among the first for the purpose named. They form dense mats of fine evergreen foliage, are uninjured by cold or heat, the flower buds are formed in the Fall, and in Spring from the middle of May to the middle of June the plants are covered with a sheet of flowers. There are several varieties, and the most common kind has dark green foliage and bright purplish crimson flowers. The variety nivalis has lighter green foliage and white flowers.

The Rev. John Nason, of Aldborough, England, has raised many fine hybrids and seedlings, and among them are three fine varieties of this phlox. Variety compacta is very dense and compact with bright rose colored flowers, variety model has very showy rosy-carmine flowers, and the bride has pure white flowers with a red center.

The perennial candytufts are also valuable. They form low spreading plants with dense and dark green foliage, and are covered the last of May and first of June with compact bunches of flowers an inch or more in diameter.

Iberis sempervirens has pure white flowers that come early. Iberis Gibraltarica has narrower leaves and later flowers that turn pinkish with age; and a new hybrid variety has red flowers, this will be, if it proves to be as represented, a valuable acquisition.

Veronica rupestris forms a mat of short trailing stems covered with fine foliage, and has, in June, abundant short spikes of dark blue flowers.

The double sneezewort, archillæa ptarmica flore pleno is valuable for a Summer and Autumn display of white. It is easily propagated and hardy, and bears its full, double, pure white flowers in the greatest profusion right up to hard frosts. The flowers are valuable for cutting, of good substance and very pretty.

A fine plant is the variegated day lily, funkia lancifolia variegata. The variegation is very marked, the whole plant appears a bright yellow in Spring and Summer, and holds its color well late into the season. The edges of the leaves are undulated, the plant is dwarf and compact, and nothing makes a more beautiful and desirable edging.

The golden leaved thyme, thymus vulgaris variegatus, is very bright in Spring, and would be valuable for beds.

There are white foliage plants that might also be used for Summer bedding; antennaria, cerastium, and artemisia, with evergreen foliage forming dense tufts.

For yellow, in Spring, the erysimum pulchellum is very fine. The plant forms oval masses of fine evergreen foliage, and is covered with the brightest yellow flowers in early Spring that last for several weeks.

For permanent beds the grasses are unexcelled. There is now such a variety with handsome plumes and veriegated foliage that a most interesting and graceful bed may be made that will be an attractive feature in any grounds.

Plants to be used for early bedding should be placed in a nursery prepared for the purpose. Small plants may be put in either in Spring or Fall, and after a season of vigorous growth will be from six to twelve inches in diameter and ready for use.

In Spring they should be taken up carefully with a ball of earth attached to the roots, and planted thickly in the beds where wanted. They will form a dense mat of green and come into flower as soon or a little later than they would if undisturbed. The flower will last in favorable weather from three to four weeks. Then, when the beds are wanted for other uses the plants may be removed, and divided if too large, and then be transplanted into the nursery. This frequent transplanting, instead of being an injury to the plants, is a benefit, when they are used for this purpose, for the roots are made compact and restain the earth, an important consideration for the success of the plant in moving.

There are many hardy sempervivums and sedums that would be of value for temporary or permanent beds. The species and varieties vary in form, color and size greatly, and many of them have pretty and showy flowers in pink, purple, white and yellow shades. They are easily propagated, can be moved safely at any time, are perfectly hardy, and a large stock could be carried with very little expense. The sedums and sempervivums and the hardy opuntias will grow anywhere they can find a little soil or a moist crevice in which to push their roots, and many barren ledges and rough rocks might be made interesting by filling crevices, pockets, and beds with these hardy succulents.

The use of hardy herbaceous plants for the decoration of public grounds is not an untried experiment. It has found able advocates among gardeners, and practical results may be seen in Swan Point Cemetery, Providence, R. I., where the hardy plants have been used extensively to great advantage by the superintendent, Mr. Timothy McCarty.

There are also many shrubs and small growing evergreens that could be used in making permanent decorative beds and groups in public or private grounds. There has been, within some years, a considerable addition to those plants that are attractive throughout the season by reason of their richly colored foliage; and there is a broad field for the gardeners who will pull out of the well worn ruts and try something new, and use to better advantage the many valuable hardy plants within their reach.

The Variegated Anthericum.

A plant may be very beautiful and yet very difficult to propagate or to rear, and, consequently, will always remain rare; such a plant can never become a popular favorite, though we may admire it. It does not readily respond to the affectionate care we may give it, and so fails to receive the hearty sympathy which many humbler plants elicit. The Daisy, the Violet, the Pansy, the Forget-me-not, and many other highly prized plants, owe their popularity not only to their beauty, but also to the ease with which they are cultivated. The Geranium, or Pelargonium, could never have become so universal a favorite if it had not been so easy to multiply and raise that even the most unskilled plant-raiser can succeed with it. It is always pleasant to notice such a plant, for one has the assurance that any one who attempts its cultivation will succeed.

The variegated Anthericum has been sent out for several years, and yet it is certain that it is not well known. It is a Liliaceous plant, and its parallel-veined gracefully curving leaves are bordered on each edge by a white stripe, making it very showy. The plant blossoms freely during Spring and Summer, throwing out numerous long, drooping flower-stems bearing small, white flowers. Each node of the flower-stem is furnished with a small bract, and at this point leaves and roots form. After blooming the flower-stems can be cut off, laid on the soil and covered a little, and in a short time the rooted plants can be cut away and potted. This Anthericum is excellent for vases, baskets and pots, growing well in the greenhouse, or window, quite free from disease, and little subject to insects. Its roots are thick and fleshy, requiring plenty of room. Another variety, known as picturatum, has a cream-colored stripe through the central portion of the leaf, while each edge has a green border. It is quite similar to the other in habit.

Hints on Window Gardening.

I have often heard the remark among professional gardeners, speaking of a certain plant not very well known to the person addressed, "give it rather a wet than a dry treatment," or vice versa, and such a hint to one having a fairly good general knowledge of the requirements of plants is usually sufficient.

Take, for instance, Cacti, Echeveria, Sempervivums, and all succulent plants, together with the Hoyas, Geraniums, Oxalises, and nearly all bulbs, and although they will all take an abundance of water when growing freely, if properly drained, yet it may be said of them in a general way that they require a treatment inclining rather to dry than wet, and will succeed in a dry atmosphere.

Again, take Ferns, Selaginellas, Pandanus, Palms, Dracænas, Fuchsias, Heliotropes, Vincar major variegata, Marantas, Caladiums, Callas, and they require rather a wet than a dry treatment, and to succeed well must have a moist atmosphere. Of course when partially resting, Fuchsias and Heliotrope require very little water and to be kept cool. Caladiums and Callas, when growth ceases, need a rest and to be kept dry. The first named should be kept warm also, not below 50° Fahrenheit. But I am speaking at present of the treatment of plants when growing, rather than resting. Moisture-loving plants would be much benefitted if a zinc tray the size of the window or table, and two or three inches deep could be used, so that fresh-growing moss could be neatly filled in between the pots and kept sprinkled more or less frequently, according to the state of the weather and the time of the year. In Spring and Fall with mild, cloudy weather, and not much stove or furnace heat in-doors, once a day would keep the moss growing. With more fire heat, or in hot Summer weather, three or four times a day would be necessary.

As to soils for window plants, the majority of those I have named, and most others, do well in the compost of leaf-mould and loam and a small quantity of well-decayed manure. For Cacti, all succulents proper, and Hoyas, I always break up some charcoal in an old mortar and mix with the soil.

I very often have would-be plant growers confess to me that they like to see beautiful flowers, and would be glad to have some always in their windows, if they would grow without much attention.

Now, what Canon Hole says to the rose grower—"He that would have beautiful roses in his garden must have beautiful roses in his heart,"—will apply to window gardeners. We must, to be successful, have them at heart. Then we shall take pains to keep informed as to their various wants. I am pretty sure that a knowledge of principles is what most persons require. This is to be obtained by reading what successful cultivators are all the time telling in the horticultural magazines.

We shall have more accounts of successes with plants, and fewer complaints of bad results when all of us endeavor to treat our window plants as fellow creatures, for so they are; and will repay richly whatever watchful care they need. From five to ten minutes every day ought to take care of a good window full of plants. Who would not be willing to give that?

Azaleas.

The Azaleas are found in this country and Asia. The favorite greenhouse sorts are varieties from the Chinese Azalea, A. Indica. The plants are raised mostly from cuttings, except new varieties, which come from hybridized seed. Though thriving best in the greenhouse, yet with attention they may be successfully raised in the house. In the window garden the Azalea should have a southern exposure, with plenty of fresh air, and not be over heated. Regular watering is one of the main conditions. It is not necessary to water every day, but never let the plants get entirely dry, especially when flowering. Daily sprinkling of the leaves is also beneficial, unless the plants show flower buds. During the time of flowering the plants should be given the coolest place, as the flowers will keep three or four days longer in a low temperature. The flowers are both single and double, and are from two to three inches in diameter, of a great variety of colors. The plant is covered with flowers from January until April. After flowering the seed-pods will commence to form, and these should be cut off, and the plant prepared for transplanting by trimming. In transplanting remove the plant from the old pot without disturbing the roots. If the soil at this time is too dry, water it thoroughly, so that the plant can be lifted with the ball of earth. One size larger pot is sufficient, and it should have some pieces of broken pots or charcoal for drainage at the bottom. Peat mixed with sand is the soil used for Azaleas. After transplanting the plant should be kept in a very cool room, but with plenty of light and sunshine. The daily sprinkling of the leaves must be resumed. Those who would like to take cuttings should improve the opportunity at the time of transplanting, and the cuttings, with a little care, can be easily rooted in sand under glass. During the Summer months, or as soon as night frosts are over, the plant in the pot may be plunged in the open ground in an airy and sunny place. Water should be given the plant as needed, and on hot days this will be at least twice, morning and evening. A few weeks before removing the plant to the house, liquid manure may be supplied twice a week.

Vases.

It is astonishing how many plants of unlike nature will grow and flourish and bud and blossom, when crowded in a vase, or given space, while plants of one kind growing in the same space would, inevitably, spindle, and droop and die. As an illustration of this peculiarity of plants I will try and tell you of an experiment of mine which resulted in the most striking success.

For a vase for my purpose I utilized a "plum-basket" which had been made of the half of a flour barrel. A row of half-inch auger holes extended around the middle of the "basket," and a dozen or more holes were in the bottom. For a pedestal for the vase I used a round block, eighteen inches high. A favorable location was chosen, and the vase mounted on its pedestal. In the bottom of the vase a lot of sticks were laid criss-cross. The vase was then filled to within four inches of the top with soil from a spent hot-bed; a lot of fresh sand had been added to this soil, and thoroughly incorporated therewith; indeed, it had been worked over and over until it was just as light and mellow as it could be. Into this mellow bed we slipped from the pots a number of plants—a miscellaneous collection taken at random from a flower-stand. In the center we planted a seedling canna, then nearly two feet high. This canna is so beautiful that it is deserving of more than a passing notice. The leaves and stalks are of rich, reddish-brown color; the leaves have a shiny, satiny surface. The plant is now budding to bloom. It is towering aloft above the others. Around this stately plant we planted three spotted-leaved callas, intermediate between the callas, dry bulbs or scarlet gladiolus were placed. Next to the edge of the vase, to trail over it, we planted a cisons discolor, a white double-flowered ivy-leaved geranium, two drooping fuchsias—one red with double white corrolla, the other single with salmon petals and red corrolla. The rest of the space was filled in with double, white balsams, a variegated-leaved geranium, a pink, and a white begonia. This begonia has shiny leaves glistening as though varnished.

These are the names of the flowers that grow in my vase, But ah! how can I ever convey to your mind an idea of the effect.

A mass of varied foliage in a state of the utmost luxuriance. Intermingled with the beautiful foliage, and in exquisite and complete harmony with it, are the gleaming white and glowing scarlet flowers. The effect is exceedingly lovely. The fuchsias have bloomed in perfection all Summer long, and are still a-blooming, and will bloom on until frost. The cissus trails to the ground. The ivy-leaved geraniums, full of bloom, are trailing apace with it, and two other branches are climbing up the canna, and other branches are reposing wherever they can find a resting place amid the plants. The white begonia is spreading its glistening leaves abroad, and rising above the leaves are innumerable pure white flowers. Did I tell you there is a blood-drop begonia there, too, with its rich, drooping, blood-red flowers. And I forgot to name an ismene calathene, which is far more thrifty in this crowded vase than its mate growing in an eight-inch pot. The balsams bloom perpetually, and seem more delicately beautiful than those grown in the border.

The vase is situated in a position where it receives the full benefit of the morning sun until ten o'clock. A tall tree shades it through the day, and the rays of the late afternoon sun find their way thereto through the flickering leaves. The plants in the vase receive a copious sprinkling morning and evening. Two or three times through the Summer a little fresh soil was sprinkled over that in the vase. That is all, not a plant in the vase withered or dried. It is replete with beauty, and I am delighted with my experiment.

Lemon Verbena.

The citron-scented plants are among the most favored and grateful to mankind, whether the Lemon Balm found in old gardens or astray by the wayside, and beloved by all the bees, Lemon Thyme choicest of borderings, Cedronella or Citronella, pleasing even in barbers' essence, the Molucca Balm or Shell Flower, Lemon Geranium, and perhaps a dozen rarer things of kindred fragrance. The lemon principle wherever found is the purest nervous stimulant, acting on the whole system through the sense of smell, as well as in refreshing waters and teas. Spanish, French, Provencal and Mexican women depend on their lemon teas as our fair English neighbors on the Chinese herb, which in novels they are represented as perpetually sipping, and really drink about five times a day. Is Dolores or Rosita upset by a neighborly affront, the ill behavior of her children, or the breaking of a string of beads, a burst of language instead of a burst of tears works up her feelings, and she flies for relief to the garden for a handful of her favorite lemon plant, steeps a pitcherfull and solaces herself at intervals the rest of the day sipping it cold. It lays a hand of calm on her grieving nerves, and she feels made anew. In passing the Lemon Verbena and catching its blissful scent, I feel as if a whiff of some life-giving elixir had been granted that would renew one if he could only catch enough of it. One wishes American women, poor, nervous, unstrung creatures, would get used to taking lemon tea, lavender water, Hungary water, or other of those pleasing cordials of flowers whose benefits our ancestors knew so well. Nerves are best treated by perfumes, which affect them as alcohol enters the blood. I have cured an oppressive headache, an imaginary ailment, by the scent of a fresh damask rose, have alleviated the nameless writhing torture of spinal ailment with cool garden perfumes, and seen racking neuralgia relieved by the odor of sweet flowers. Why not? We know that faintness, vertigo and nausea, symptoms of extremity, can be driven off by lavender, citron and camphor scents, and sensitive persons have died of passing through vile odors; why should not these things be owned as agents of no common strength, and used for all they are worth? Sweet odors quiet nerves. Nothing known to American gardens is more a fountain of pure scent than the Lemon Verbena.

You may fancy that I have a partiality for it, and guess right. Dozens of the plants flourish in my garden, and I would like to see a bush in every woman's flower border. Besides the refreshing scent, which is reviving as a whiff of camphor or ammonia, a leaf put in the tea-pot with the Chinese herb gives a fine flavor to common grades of tea. A growing plant of this Verbena, or of Lavender, purifies the air in a sitting or sick-room by chemical process, neutralizing bad air as really as carbolic acid or copperas by far pleasanter means. In old times strong herbs and their odors were the only disinfectants known, and were held proof even against the plague itself. Windows full of sweet smelling plants keep out flies, and the Lemon Verbena so fills the air about it with its vapory scent as to be specially useful in this way.

The seed is troublesome to start, being fine as Portulaca and more sensitive. To be sure of its germination it is best, like all other fine herbs, sown in the Fall, or as soon as ripe, in very fine, moist sand, with one-third or less wood-mold sifted through, and a wet flannel or felt laid over the surface, with no other covering to the seed. Or, sow it with the thinned sifting of wood-mold, and cover the box with oiled paper tacked tightly over it, giving air every few days till the plants come up. Moisten by setting the box in water. By far the best way to supply the garden is to get a plant or two, which costs no more than a packet of seed, and when it is well grown take cuttings from it, which start readily in sandy mold when covered with glass or paper. They may be rooted from June to September out of doors, or any time in a warm room. Take tender shoots three inches long, not woody ones, for cuttings.

I find Lemon Verbena grows rapidly in a rich light soil of garden loam, old sifted manure and sand. Dig out a basin fifteen inches across and as deep, in the border, put in five inches of old leaves and manure with sand over. Set the plant deep, pour a quart of water round it, and sift the prepared soil into the water without pressing. As it settles it will pack the roots just right. Turn a peach basket or light cover over the plant, and shade till it throws out new shoots. It is rather a delicate thing in some respects, not liking hot sun while it is young, and shrinking at the first touch of frost. On this account it is convenient to grow a few plants in tile pans or boxes eight inches deep, that can be moved as desirable. In the strong sandy soil it loves, with good drainage, it is hardly possible to give the plant too much water from the watering pot. Three times a day in hot, dry weather I have watered my plants, and they throve gratefully. As it is grown for leaves, not for its pale violet flowers, it is best to clip these as soon as they appear. Nipping the buds only makes them start twice as many blossoms later, so let them flower and cut the bloom at once, before the plant loses force in the fertilizing process. Then give some guano water, or lawn dressing, to have it make fresh sprays. The sprigs may oc cut for drying at this time, if wanted, and are fragrant to the very stems as long as they last. But it is best to keep the Lemon Verbena as a window plant to perfume the room through Winter; the greatest care it needs being to guard from frost and sudden changes of temperature. A plant lives twenty years if not frosted, and grows into a lovely shrub many feet high, its willowy, pale green foliage having much the grace of the California Pepper tree or Australian Acacia. Lemon Verbena in Pacific gardens grows in a season or two to the size of a large Castor Oil plant, the rich volcanic sand and even temperature suiting its refined senses to a charm. A well grown plant is a fine ornament in sitting-room or conservatory, especially contrasted with the dark leaves of a Myrtle or Camellia.

Echeverias.

The Echeverias were brought prominently into cultivation a few years since for the valuable purpose they served in carpet bedding, and it is for this use they are now mostly in demand. They are fleshy, succulent plants, closely resembling the Sempervivums, and their general treatment in cultivation is the same.

Some of the species, however, have been considered sufficiently ornamental to be worthy of a place in the greenhouse among Winter-blooming plants. Conspicuous among them are E. secunda and some of its varieties, especially glauca. The cut flower stems are very graceful with other flowers in vases, and are quite lasting. The flowers are of a neat, regular form, wax-like in appearance, with shades of orange and orange-red. The plants are easily grown in good loam in ordinary greenhouse temperature. They are propagated by the small offsetts at the base of the rosette which, when removed, root easily.

Plants intended for Winter blooming should be healthy specimens in early Spring, potted in good loam; keep them growing through the Summer in a light, airy place, and remove any flower-stems that may appear during the Summer—with the arrival of Autumn this care may cease, though by removing them still later flowering may be further retarded.

In fine bedding work the Echeverias can can be employed very effectively.

Salpiglossis.

The great variety of quaint-colored flowers amongst the Salpiglossis has made them special favorites of mine, as I find them to be very attractive when tastefully arranged in vases of suitable patterns and dimensions. I have used these flowers rather freely during the Summer with sprays of sweet briar as greenery, and although they do not last long in a cut state, when lightly arranged they are sure to be admired. I cut them with stems of various lengths, and select the lightest and brightest colors for the center of the vase, with the dark ones round them. When arranging them I never crowd the flowers. Many people use too many flowers in their arrangements. As a consequence they are so crowded that they never look well. In many cases if only a third of the quantity were used the arrangements would be more effective. I fill my vases three parts full with clean silver sand, and keep it regularly moist by pouring a little water over it every other day. I prefer sand to filling the receptacles with water only, as the sand enables one to keep the flowers in an erect position, as it is not pleasing to see them tumbling over the sides of the vase.

Dicentra Canadensis.

Among our many early Spring flowers there is none prettier or more graceful than Dicentra Canadensis. It grows in rich woods in nearly all the Northern States and as far south as Kentucky. Its finely cut, fern-like leaves of delicate green, and its racemes of oddly shaped, nodding flowers, make it very attractive. The corolla is heart-shaped, and of a greenish-white color tinged with rose, and the flowers have a fragrance somewhat like that of the hyacinth. The roots consist of small tubers, yellow in color, and resembling grains of Indian corn, which fact has given it the common name of squirrel corn.

It is easily transplanted and takes kindly to cultivation, though it does not blossom quite as freely in the garden as in its native woods.

www.ingramcontent.com/pod-product-compliance
Lightning Source LLC
Chambersburg PA
CBHW032244080426
42735CB00008B/993